WHO CONTROLS PUBLIC LANDS?

The University of North Carolina Press

Chapel Hill and London

CHRISTOPHER McGRORY KLYZA

MINING,

FORESTRY,

AND

GRAZING

POLICIES,

1870–1990

WHO CONTROLS PUBLIC LANDS?

© 1996 The University of North Carolina Press
Manufactured in the United States of America

The paper in this book meets the guidelines for
permanence and durability of the Committee on
Production Guidelines for Book Longevity of the
Council on Library Resources.

Library of Congress Cataloging-in-Publication Data
Klyza, Christopher McGrory
Who controls public lands?: mining, forestry,
and grazing policies, 1870–1990 /
Christopher McGrory Klyza.
p.   cm.   Includes bibliographical references
(p.) and index.
ISBN 0-8078-2264-7 (cloth: alk. paper).
ISBN 0-8078-4567-1 (pbk.: alk. paper)
1. Public lands—Government policy—United
States—History. 2. Mineral lands—Government
policy—United States—History. 3. Forest policy
— United States—History. 4. Range policy—United
States—History. 5. Public lands—United States
—Management—History. I. Title.
HD216.K55   1996   95-23352
333.1'0973—dc20   CIP

00 99 98 97 96   5 4 3 2 1

CONTENTS

TABLES

PREFACE

Having grown up in the East, I thought of the public lands as places like the Grand Canyon, Yellowstone, and Yosemite, places of extraordinary beauty located in the West. As I learned more about them—and I realized that these lands were more than national parks—I became more intrigued. My interest in these lands crystallized the summer after graduating from college. Before starting graduate school in natural resources policy, I took a two-month drive through the western United States and Canada. Indeed, in some ways my drive was not unlike the one described in "Conservation Esthetic," an essay by Aldo Leopold: "Everywhere is the unspecialized motorist whose recreation is mileage, who has run the gamut of the National Parks in one summer." However, that drive gave me a focus for my graduate work, a focus that continues to this day. My interest began with a question: Why is it that in the nation most committed ideologically and empirically to free-market capitalism, the federal government owns nearly one-third of the land? As I read and learned more about the public lands, I became fascinated with the complicated history of public-lands policy and the myriad of agencies devised to manage these lands. The public lands have played a significant role in the history of the United States, especially in the West, and I think they will continue to play a crucial role in our society. It is on these lands that we must find a new balance in the relationship between humans and nature. It is here that we have the best chance of maintaining and restoring the biological diversity of the country. If as a society we fail on these public lands, it is hard to be optimistic about our ability to succeed on private lands. I still live in the East (unlike most analysts of the public lands), and I have learned that the public lands are not exclusively a western phenomenon. In fact, I can see public land—the Green Mountain National Forest—from the windows of my home in Bristol, Vermont.

This book is a work of synthesis and reinterpretation. I have not discovered a new collection of papers by Gifford Pinchot, nor will I argue for the newfound importance of some obscure statute. Rather, I rely on secondary sources and government documents to retell the story of public-lands policy, with contemporary work in the social sciences as the basis for this retelling. So, this is a work of historical political science, not political history. I argue that distinct policy patterns exist for the management of mining, forestry, and grazing on the public lands, and that the institutionalization of a specific idea at the time the federal government started to manage each of these resources is the basis for these different patterns. Different ideas were institutionalized and became embedded in each policy regime, leading to different policy patterns. Many current analysts of public-lands politics argue that each of these policy areas today is dominated by commodity interests—mining companies, forest-products corporations, and ranchers. That may be true, but they have arrived at this end through very different means, which implies that they might depart from current policy in significantly different ways as well.

This book is the culmination of many years of schooling, thinking, research, writing, and revising. My thinking on natural-resources policy has been informed and shaped by the faculty and students in the Department of Natural Resources at Cornell University, the School of Forestry and Environmental Studies at Duke University, the Department of Political Science at the University of Minnesota, and the Departments of Political Science and Environmental Studies Programs at the University of Vermont and Middlebury College. More specific thanks are due to my dissertation committee at Minnesota—Terry Ball, John Freeman, Virginia Gray, C. Ford Runge, and Rod Squires; fellow graduate students at Minnesota—Mike Barnett, Larry Biskowski, Kurt Burch, and especially Eric Mlyn and David Sousa; colleagues at Vermont—Eileen Burgin, Pat Neal, and Bob Pepperman Taylor; and colleagues at Middlebury—John Elder, Russ Leng, Alison Stanger, Alec Stone, and Steve Trombulak. I also thank Middlebury College for a leave that allowed me time to complete the project. At the University of North Carolina Press, I offer thanks to my editor Lewis Bateman for his support of this project, and to the design and technical staff for their efforts in publishing this book. Special thanks go to Sheila McGrory-Klyza, who has read over portions of this manuscript with a keen eye, encouraged me through completion of the manuscript, and made the final years of work much more meaningful. Lastly, a sincere thank-you to my late parents, who provided tremendous support throughout this project as well as my entire education.

Some arguments and factual material in this book originally appeared in the following publications: "A Window of Autonomy: State Autonomy and the Forest Service in the Early 1900s," *Polity* 25 (Winter 1992); and "Ideas, Institutions, and Policy Patterns: Hardrock Mining, Forestry, and Grazing Policy on United States Public Lands, 1870–1985," *Studies in American Political Development* 8 (Fall 1994). I am grateful to these journals for permission to reprint this material.

PUBLIC-LANDS
PUZZLES

With the election of Bill Clinton and the appointment of Bruce Babbitt as secretary of the interior, environmentalists were confident that reforms in public-lands management would soon be adopted. Indeed, three such reforms were included in President Clinton's first budget: increasing grazing fees on public lands, charging royalties for hard-rock minerals removed from public lands, and eliminating timber harvesting on the national forests when the government looses money (referred to as below-cost sales). The administration justified these changes by asserting that they would reduce the budget deficit, but its main objective was to change public-lands management. Within a month, all three proposals were stripped from the budget due to strong pressure from western members of Congress. This was a rude awakening for the administration, and it has been enmeshed in public-lands politics ever since. Let's take a closer look at what has happened in the issue of grazing reform.

Grazing policy on public lands managed by the Bureau of Land Management (BLM) and the Forest Service has been contentious since the 1970s. Environmental groups have argued that ranchers have too much control of these lands, that the lands are overgrazed, leading to reduced forage for wildlife and environmental damage, and that the fees charged ranchers for grazing permits are far too low. In 1993, the grazing fee was $1.86 per animal unit month (AUM), the amount of forage needed for one cow and a calf or five sheep for one month. Grazing fees on private lands averaged more than $8.00 per AUM, though federal permits had greater maintenance costs. Less than 28,000 permittees make use of federal grazing permits on more than 250 million acres throughout the West.

In Clinton's original fiscal year 1994 budget, he proposed raising grazing fees to $5.00 per AUM. This led to strong protests from western senators. The fees survived in a Senate vote, but then a group of western Democratic senators, led by Max Baucus (Mont.), told Clinton their support of his economic package was wavering in the face of these increased grazing (and mining) fees. In response to this pressure, Clinton removed the increases from the budget and pledged to reintroduce them later. This suggested that the Clinton administration was not willing to spend its precious political capital on grazing fees, and that it had underestimated the difficulty of public-lands politics.[1]

By the summer of 1993, Babbitt had already begun to alter his approach to reform, seeking to involve the ranching community and western politicians rather than to dictate reform to them. An environmental-impact statement was drafted to accompany proposed executive-branch reforms on grazing fees, rangeland condition standards, and the general administration of the federal grazing program. The public was invited to comment, and a series of open meetings were held throughout the West in the spring and summer. On July 30, a group of eight western senators introduced their own grazing-reform legislation, seeking to develop an alternative preferable to Babbitt's and the environmentalists'. The proposal included a new fee formula, one that would increase grazing fees 25 percent. Wyoming senator Malcolm Wallop (R), one of the cosponsors, said he was happy with the current fee formula but recognized that some change would probably be forthcoming. The new bill would better protect ranching interests, whereas the Babbitt package was part of "a vengeful assault by people who know nothing about the industry."[2]

Trying to keep the pressure on Congress, the Clinton administration announced on August 9 that it would administratively more than double grazing fees, from $1.86 to $4.28 per AUM, by 1996. In addition, the length of the grazing season would be shortened, use of pesticides terminated, water rights of ranchers on the leased lands curtailed, conservation of riparian areas stressed, and leases broken if ranchers mismanaged their lands. In announcing the action, Secretary Babbitt indicated that these changes were the beginning of major reform in public-lands management. The changes represent "our commitment and our responsibility to live more lightly on the land," the secretary said. Because Congress had passed his budget the week before, Clinton felt he could now return to this sensitive issue without the risk of losing votes. The action was strongly criticized by western Republican senators, but only mildly criticized by Senate Democrats from the West.[3]

Opponents of the fee increase did not acquiesce, however. On September 14, the Senate voted fifty-nine to forty to approve an amendment to the Interior Department appropriations bill that would prevent the proposed increase of grazing fees and changes in land management for at least a year. After much debate in the conference committee, a deal was struck by Babbitt, House members, and Senate Democrats on the committee. The changes in grazing management and fee increase were retained, though the increase dropped from $4.28 to $3.45 per AUM. Western Senate Republicans continued their strong opposition to any of Babbitt's proposals. Pete Domenici (N.Mex.) said he feared the increases would "result in unnecessary pain and economic suffering for thousands of New Mexico's ranchers and small townspeople." The executive director of the New Mexico Cattle Growers' Association commented that the government was "in the process of destroying the lives of thousands of innocent people."[4]

Despite the loss in conference, the ranchers and their allies still were unwilling to concede to the increases. Although the House passed the conference bill by a three-to-one margin, when the Senate began consideration, western senators filibustered to block a vote. The Senate vote to end the filibuster on October 21 was seven shy of cloture. There were two further votes to end the filibuster, but no more than fifty-four of the needed sixty senators ever voted to end the debate. According to Senator Alan Simpson (R, Wyo.), the debate went beyond grazing: "We are defending a Western life style in this Administration's war on the West." Babbitt vowed that if Congress did not act, he would make the changes in grazing policy by administrative order.[5]

On November 9, Senator Harry Reid (D, Nev.), engineer of the grazing compromise in the conference committee, agreed to drop all grazing language from the appropriations bill, recognizing that the filibuster could not be broken. Both sides claimed a limited victory. Ranching interests and their supporters were able to prevent Congress from passing a law raising grazing fees and making land management changes. Babbitt and environmentalists claimed that Congress would not now block his administrative proposals made in August to raise fees and improve management of grazing lands. Indeed, fees could now be raised higher. It was clear to all, however, that Babbitt and the environmentalists did not have the power to overcome the western senators on this issue.[6]

In 1994, Babbitt continued his effort to administratively raise grazing fees and make additional changes in grazing management. But in recognition of the power and opposition of the ranching interests, he pledged to create new multiple-resource advisory panels in each state to help determine how the regulations would go into effect. These new councils would give local people, rather than Washington, increased influence in administering the regulations. Furthermore, in February Babbitt forced BLM director Jim Baca to resign. Baca had been strongly supported by environmentalists and had fiercely fought for the changes in grazing policy. Babbitt found his style too confrontational for this new approach, and, after being pressured by western politicians, asked for his resignation. These moves, and Babbitt's frequent meetings with ranchers throughout the West, were criticized by environmentalists, while ranchers remained skeptical.[7]

The November 1994 elections all but put an end to Babbitt and Clinton's efforts to reform public-lands grazing policy. With Republicans gaining control of both the House and Senate, Babbitt announced in December that grazing fee increases would be left up to the new Congress. He also delayed the remaining grazing regulations for six months to allow Congress time to vote on them. Most analysts thought the proposed changes were dead.[8]

So, after two years of effort, the grazing fees remained where they were and management of the grazing lands did not change. This fierce political battle also spilled over into other political arenas. Babbitt was a leading candidate for two openings on the Supreme Court. When he was mentioned for the slot in 1993, environmentalists fought hard to keep him in the cabinet because they thought he was so good. When mentioned for the 1994 opening, the environmental movement was not particularly vehement about keeping him in Interior. His problem this time was opposition from many western Republicans in the Senate, who were opposed to his grazing politics and other environmental positions. Overall, then, the raising of grazing fees proved far harder than Babbitt, Clinton, and others had imagined. Indeed, the battle over this seemingly obscure issue contributed to the view that Clinton's presidency was adrift.

This story has two chief lessons. First, groups are very powerful in the American political system, but, as you will also understand by the time you finish this book, ideas and administrative capacities are also crucial to understanding public-lands politics. And second, any efforts to significantly alter public-lands management must pay attention to the history of past policy, for it is in this history that the roles of groups, ideas, and administrative capacities develop specific patterns. This book traces the history of hard-rock mining policy, forestry policy, and grazing policy on the public lands since the inception of these policies, from more than one hundred to sixty years ago, and will thus give the reader an understanding of how these patterns developed and why they are so difficult to change.

The federal government owns 662 million acres of land, 29 percent of the land in the country, acquired through conquest, purchase, and treaty. The vast majority of this land is managed by four agencies: the Bureau of Land Management (269 million acres), the Fish and Wildlife Service (89 million acres), the Forest Service (187 million acres), and the National Park Service (80 million acres). Federal ownership is concentrated in the western states, ranging from 82 percent ownership in Nevada and 68 percent ownership in Alaska to .4 percent ownership in Connecticut and Iowa. These lands are significant sources of commodities such as forage, minerals, oil and gas, and timber; serve as important sites for outdoor recreation; and serve as crucial areas for wildlife habitat and general land preservation.[9]

From the mid-1800s to the mid-1930s, the federal government initiated programs to manage three types of resources on these public lands. The discovery of gold in California and elsewhere in the West demanded some government policy, because the minerals were on its land. The government responded in the 1860s. Debate over the nation's forests began in the 1870s, and a system of national forests to be managed by the federal Forest Service was created in the late 1800s and early 1900s. And in the 1930s, the government finally began to manage grazing on lands outside of the national forests.

One would expect the management patterns for these three resources on the public lands to be very similar. After all, the resources are all located on land controlled by one entity, similar governmental agencies administer these lands, and the actors interested in how the lands are managed share many similarities. Yet a comparison of these three policy regimes reveals three significantly different policy patterns since their inception.[10] In mining policy, the pattern is *privatized*. The federal government allows private firms to remove minerals from public lands for minimal fees or transfers mineral-bearing public lands to the private sector for $2.50 or $5.00 per acre. In forestry policy, the pattern is *professional*. Congress created an agency of professional foresters and charged it with managing the national forests. These foresters follow a clear and coherent professional code in making their decisions. And in grazing policy, the pattern is *captured*. Although Congress created an agency to administer private use of these lands, this agency and its decisions have been dominated by the cattle ranchers that use the land. What accounts for this puzzle? That is, what explains different policy patterns in the same policy area?[11]

Previous studies of public-lands politics have not addressed this problem adequately for four main reasons. First, some studies have been descriptive without being theoretical (i.e., they have not explained what has happened). For instance, Charles Wilkinson offers a fine and useful introduction to public-lands politics, stressing the importance of a historical approach to understanding current politics. But that is as far he goes; he does not explain how this history has led to different policy patterns. Indeed, he suggests that public-lands politics is all of the same type: captured. As I will demonstrate, this is not the case. Another problem is that some studies have made insufficient comparisons across policy regimes. David Clary's study of the Forest Service and Robert Durant's study of the BLM are fine individual case studies, but by focusing on only one policy regime their findings are not illuminated by the contrasts and commonalities found in other cases. For example, Clary concludes that the Forest Service is a captured agency, but is it when compared with the BLM or mining policy? A third problem is that some of these studies have focused on only specific agencies rather than on the entire policy process. Paul Culhane's comparison of the BLM and the Forest Service and Jeanne Clarke and Daniel McCool's examination of seven natural-resources management agencies come to different conclusions. Culhane reports that the BLM and the Forest Service are becoming more alike, primarily because the BLM is becoming more professional, like the Forest Service. Clarke and McCool report that the Forest Service remains a vastly stronger agency than the BLM. Both of these studies focus on the administrative capacities of agencies, failing to examine the larger policy regime and to offer explanations of the patterns of public-lands politics. For instance, Clarke and McCool overstate Forest Service strength and flexibility and ignore its declining public stature, internal disarray, and less dominant position in the forestry policy regime. The final problem with many such studies is their lack of a systematic historical perspective, most clearly seen in Culhane and Durant.[12]

As will be borne out in what follows, such a historical, cross–policy regime approach to examining public-lands politics is crucial to understanding current policy patterns and determining how such patterns might change. By understanding that the idea of interest-group liberalism was embedded into grazing policy when it was formulated in 1934, and that interest-group relationships and administrative capacity have revolved around this idea, Clinton and Babbitt would have understood that changing grazing policy would be no small feat. Rather, they would know that the grazing policy regime would have to be shaken at its very foundations in order to achieve significant change.

The key, I think, to understanding the puzzle of different policy patterns in public-lands politics is in understanding the foundation of a policy regime and the subsequent politics that emerge from this. I argue that (1) at the inception or foundation of a new policy regime, one particular idea becomes embedded in the state, an idea reflecting the prevailing views of the appropriate role of government at that moment; (2) this idea becomes a privileged idea that serves to guide and constrain the state actor(s) within this policy regime; and (3) once institutionalized, this idea becomes very difficult to dislodge, despite challenges from interest groups and agencies supporting other ideas.[13] This institutionalization, though, is not forever. In the face of repeated challenges and a changing society, cracks in the foundation of institutionalization occur and policies based on nonprivileged ideas can be victorious. Cracks beget cracks, and at some point, the idea can fall from its privileged position. If this happens, it is likely that the politics within the policy regime will become more open, as the different societal ideas are considered in a more "equal" setting. It should also be stressed that ideas do not determine all that happens. The importance of ideas is constrained by groups and institutions, just as groups and institutions are shaped by ideas.[14]

The inception of any new policy regime is a key time. When the government first becomes involved in a particular area, it establishes new institutions or institutional responsibilities and establishes a set of expectations about how the state and society will interact in this policy area.[15] Because of this, how the government undertakes policy at this beginning serves to channel or guide future policy choices and directions. At this time, ideas play an important role, because state and society are engaging in a new venture, one that by definition they have not attempted before. Hence, ideas play an important role in supplying information to the government as to how it should proceed.

Once an idea is adopted, it is likely to become embedded within the state through the design of new institutions and the establishment of norms and procedures in the particular policy regime. This idea becomes privileged within the universe of ideas in society that deal with the particular policy regime. This privileged idea does not prevent the rise of other ideas, nor dictate the role of groups in politics. It does, however, help to channel the future of the policy regime, to determine the contours of the policy debate, and to determine the boundaries of policy alternatives.

Change, obviously, does occur. These embedded, privileged ideas can become dislodged. Societal challenges can arise (for instance, social movements in favor of women's rights and in opposition to the policy status quo), the context in which the idea became embedded can change (for example, the collapse of the Soviet Union), or the institutions in which the idea is embedded can change due to the rise of a new profession, reorganization, or an expansion in agency responsibilities. Any and all of these factors can lead to changes. Nevertheless, when ideas are embedded in the state, these changes do not come easily. Privileged ideas are difficult to dislodge, a lesson Clinton and Babbitt learned in grazing policy. This is because these ideas are a force of friction, a source of stickiness, in this process of policy change.

This focus on ideas is not meant to undermine the importance of groups or agencies in public-lands politics. As will be apparent in the following chapters, groups play a very important role in these cases. The point is that the arena in which these groups interact and how they interact is greatly influenced by these ideas embedded within policy regimes. These ideas are important for group cohesion and group strategy, something that is missing from accounts that focus solely on groups and ignore ideas. Similarly, the roles of agencies and their administrative capacities are also determined in part by the idea that undergirds the policy regime.

In the following chapters, I will (1) examine the three main ideas that have served to shape public-lands politics since the late 1800s; (2) demonstrate how one of these ideas became embedded within the mining, forestry, and grazing policy regimes, guided policy, and has fared in response to agency and interest-group challenges based on other ideas; and (3) analyze the cases to suggest how ideas, groups, and the state interplay in a manner that can enrich our understanding of public-lands policy.

In Chapter 2, I analyze the three main ideas that have been present in the public-lands policy debate since the late 1800s: economic liberalism, technocratic utilitarianism, and preservationism. In examining these ideas, I will explain the core of the idea, discuss where the idea came from, and trace the evolution of the idea from its inception through the present.

Within each of the three general case-study chapters (dealing with mining, forestry, and grazing), I examine three crucial episodes in the development of the policy regime. In all three policy regimes, the first episode examined is the inception of government management of that particular resource on the public lands. The two remaining episodes are selected to demonstrate both the existence of a policy pattern and to demonstrate how the privileged idea fares in light of significant challenges based on other ideas.

The subject of Chapter 3 is hard-rock mining policy. The specific cases examined are the establishment of the mining policy regime in the wake of the mineral rushes from the 1840s to the 1860s (the mining laws of 1866, 1870, and 1872), the continued allowance of mineral exploration and development on wilderness lands (the Wilderness Act of 1964), and the debate over access to the public lands to search for strategic minerals in the 1970s and 1980s. The focus in Chapter 4 is forestry policy. I examine the formation of the forestry policy regime in the 1890s and 1900s, the passage of the Wilderness Act of 1964, and the failure of the privatization initiative of the 1980s. And in Chapter 5, I examine grazing policy. The cases I focus on are the inception of the grazing policy regime through passage of the Taylor Grazing Act of 1934, the passage of the Federal Land Policy and Management Act of 1976, and the controversy over grazing fees that flared in the 1970s and 1980s.

In the final chapter, I compare the findings from the three case-study chapters to illustrate why different patterns developed. I next discuss the current tensions within economic liberalism, technocratic utilitarianism, and preservationism and examine the present status of mining, forestry, and grazing policy. Based upon these findings, I close by speculating on the future of public-lands politics, with special attention to the importance of ideas and institutions.

THE IDEAS

COMPETING CONCEPTIONS OF THE PUBLIC INTEREST

Since the late 1800s, three main conceptions of the public interest regarding public-lands management have been dominant: economic liberalism, technocratic utilitarianism, and preservationism. Each of these ideas has played an important role in shaping public-lands politics, specifically in mining, forestry, and grazing. As discussed in the chapters that follow, one of these ideas was embedded within the state at the inception of the policy regimes. This embedded idea became institutionalized within the state, gaining a privilege that helped to determine the policy pattern within the policy regime. Additionally, supporters of the other ideas would challenge this privileged idea throughout the history of the policy regime. In this chapter, I explain each of the three ideas and trace their historical paths since the late 1800s.

Economic Liberalism

In the period following the Civil War through the early 1900s, economic liberalism was the dominant idea regarding economic activity. Although government action was supported to help "the release of energy" within society, it was believed that minimal government intervention consistent with this need best served the public interest. This perspective opposed government ownership of the public lands, even private ownership with government regulation, for this, too, was viewed as a hindrance to the public interest.[1]

Since the early 1900s, a variant of the economic liberalism perspective has developed that favors the transfer of the public lands to the individual states. This variant holds that the states are closer to the people and commerce. Usually implicit in this variant is the idea that state governments will transfer significant amounts of the public lands to the private sector. The interest-group liberalism that emerged in the 1930s promoted continued federal government ownership because of the system of small private-group control favored by this outlook that had developed in the management of these lands.[2]

Among the leading advocates of private ownership of the public lands around the turn of the century were the states, because they felt it would help development, increase taxes, and provide for more self-government. Many members of Congress and the executive branch also were opposed to the idea of permanent public lands, arguing that it ran counter to the free-enterprise system. For example, in response to a Theodore Roosevelt initiative to establish a coal-leasing program for the public lands, the minority report of the House Committee on the Public Lands opposed the plan as "dangerously socialistic, paternalistic, and centralizing in its character" and "not in harmony with our institutions, with our past policy and practice, or with that of any great commercial nation." Secretary of the Interior Ballinger, in office under President Taft, clearly favored private ownership as well: "You chaps who are in favor of this conservation program are all wrong. In my opinion, the proper course to take with regard to this [public domain] is to divide it up among the big corporations and the people who know how to make money out of it and let the people at large get the benefits of the circulation of the money."[3]

The advocacy of private ownership also focused on the threat of federal regulation of forestry on private land. As early as 1883 the timber industry was on record opposing the regulation of private forestry. In 1919, the Forest Service and the Society of American Foresters began a campaign to establish federal regulation of forestry on private lands. The industry opposed such regulation, calling it "un-American and unconstitutional." The timber industry also began to speak of the virtues of private ownership versus the alternative of public ownership following World War II. From 1945 through 1953, the Forest Industries Council and the National Lumber Manufacturers Association made a series of statements in favor of private ownership of land for forestry purposes and advocated the transfer of existing public forest lands to private ownership.[4]

Grazing interests led the push for moving the public lands to the private sector in the 1940s, a movement called the "great land grab." Ranchers argued that the public interest would be best served if those who used these lands owned them, and could thus be more responsive to the free market. In 1947, the governor of Colorado wrote that "after considering the public welfare, such of the public domain as properly may be eligible, and this should be considerable acreage, should be channeled into private ownership." This effort was unsuccessful.[5]

The idea of transferring public lands to the private sector arose again in the late 1960s, during the existence of the Public Land Law Review Commission. This congressional commission, headed by Colorado representative Wayne Aspinall, favored transferring significant amounts of federal land to the private sector (or, at the least, to state government). Aspinall said: "We must find the means to provide for the transfer of the public land into non-federal ownership." These ideas carried over to the Sagebrush Rebellion of the 1970s and the privatization movement of the 1980s.[6]

The privatization movement was launched by a group of economists advocating the transfer of the public lands to the private sector to increase economic efficiency. Because some of these economists were members of the Reagan administration, and because the idea fit with the administration's free-market ideological perspective, this privatization movement was seriously considered by the White House. A discussion of the specific positions of the privatization advocates and the outcome of the movement is the topic of a case study in Chapter 4.[7]

Shifting to the variant of economic liberalism, the idea of transferring the remaining public lands to the states became popular once the retention of the public lands began in the 1890s. At the Second National Conservation Congress in 1911, a group of western governors spoke in opposition of the federal conservation movement, and asserted "the superior efficiency of state management." In 1913 and 1914, strong demands for the cession of remaining public domain lands to the states were heard. At the 1914 Western States Governors Conference, the governors sought the transfer of all the public lands to the states. For them, increased private ownership meant increased taxing power for the states.[8]

In the 1920s, the forest industry demonstrated its preference for state control over federal control. Throughout the debate over the regulation of forestry on private lands, the industry opposed the regulation outright, but if it were to come, preferred state to federal regulation. In 1929, the Hoover administration supported a plan to transfer all unreserved federal land to the states (with the idea that much of these lands would eventually be transferred to the private sector). The western states, though, were not interested in the transfer because the federal government was to retain mineral rights to the land.[9]

During the effort to move lands to the private sector in the late 1940s and early 1950s, some advocated the compromise of transferring the lands to the states, but this idea was lost in the debate over transfer to the private sector. It was not until the Sagebrush Rebellion of the late 1970s that the idea of transferring the public lands to the states received much attention again. The Sagebrush Rebellion began in the 1970s when Nevada and Arizona requested that the federal government grant them more of the federal lands within their borders. In a short time, the rebellion had spread across the western states; five states passed laws claiming title to the federal lands within their borders. The case for this transfer was based on spurring economic growth and states' rights. The rebellion received national attention, and in the 1980 presidential campaign, Ronald Reagan proclaimed that he, too, was a Sagebrush rebel. The rebellion was defused in two ways with the election of Reagan: the initiation of federal policies more sympathetic to the western states and the development of the privatization initiative.[10]

A relative of economic liberalism, interest-group liberalism also plays an important role in public-lands policy. This idea of interest-group liberalism—of allowing particular interests to colonize particular state institutions—emerged on the national scene in the 1930s. This new idea of how to service societal interests has been defined by Theodore Lowi as an "amalgam of capitalism, statism, and pluralism." It is based on a system of self-government in which economic interests, organized in groups, are delegated authority over policy making in their policy realm (e.g., agriculture policy). Grant McConnell referred to the idea as voluntary cooperation, and argued that it was based on the private expropriation of public policy justified by "the orthodoxy that organization by small units is the essence of democracy."[11]

Today, some actors, especially commodity users and their supporters, and analysts argue that the current system, with its assorted subsidies to mining, timber, and grazing interests, should be maintained because it helps business and therefore serves the overall public interest (recall the opening story of the Clinton administration's attempt to raise grazing fees). The wise use movement, representing commodity interests and their supporters, is the political manifestation of interest-group liberalism on the public lands. The movement argues for a continuation of, and improvement of, policies favoring the use of public-lands resources. Analysts argue that existing long-term political and economic systems would be disrupted by fundamentally changing the management policies on these lands or by transferring these lands to private ownership.[12]

Economic liberalism has been a lasting and important perspective of the public interest regarding the management of the public lands since the federal government's retention of these lands began. At first, economic liberalism solely stressed the importance of private ownership. It later broadened to include favoring the transfer of the public lands to the states (often as an intermediary step on the path to private ownership). In addition, an offshoot of economic liberalism, interest-group liberalism, has taken on increased importance since the 1930s.

Technocratic Utilitarianism

Technocratic utilitarianism is the second major idea concerning the public lands, and it is most closely associated with the profession of forestry and the Forest Service. According to this idea, forest resources, where possible, should be retained in federal government ownership, where they could be managed by the professional foresters employed by the government to achieve the public interest. The technocratic component of this idea, which embraced state control by foresters, came from Germany; the utilitarian component, which helped determine the public interest, came from England.

The forestry profession in the United States was born in the late nineteenth century. During this period, German-born Bernard Fernow and European-trained Gifford Pinchot helped to establish forestry in the United States; prior to this time, it did not exist. The central concept they brought from Germany, where forestry was strongly established, was sustained yield: the forests should be managed for a constant, sustainable amount of wood. In addition, they had personal experience with the German policy of having the forests under the control of the central state.[13]

It is from these German roots that the technocratic aspects of the forestry profession come, and the support for government ownership of the forests. In an 1895 speech to the American Association for the Advancement of Science, Fernow stated that the forest resource "calls preeminently for the exercise of the providential functions of the state to counteract the destructive tendencies of private exploitation." According to R. W. Behan, "Fernow called here [the United States] for public ownership of an almost holy resource (which it was in Europe) to be managed by an almost holy man, the forester." He continues by arguing that both Fernow and Pinchot agreed that "private enterprise and the profit motive could not be trusted with so valuable a commodity as the forest resource, and no one but a professional forester could be entrusted with its care." Numerous studies have concurred on this technocratic nature of the early Forest Service, focusing on an internalized, shared value orientation that embraced centralized authority, efficiency, expertise, rationality, scientific elitism, and technocracy. Politicians and the public were not to be trusted with these decisions; they were not knowledgeable enough and were often swayed by special interest pressures.[14]

Pinchot, who headed the federal forestry agency from 1898 to 1910, was the foremost proponent of forestry and the idea of technocratic utilitarianism in the United States. He regarded it as the duty of the public servant to do all that was possible to serve the public good, and it was the public servant who could manage these forest resources in a fashion to serve the public good, something the private sector could not do: "Their [public servants'] care for our forests, waters, lands, and minerals is often the only thing that stands between the public good and the something-for-nothing men, who, like the daughters of the house leech, are forever crying, 'Give, Give.'" Pinchot goes on to cite with satisfaction that the Forest Service had been the most attacked government agency in the country. The more successful the Forest Service had been in protecting the public interest, the more it had been attacked by special interests. Elsewhere he claimed that proper management of the forests was best achieved under state ownership: "Forest property is safest under the supervision of some imperishable guardian; or, in other words, of the State."[15]

Later, when government regulation of private forests became an issue, Pinchot again supported federal control. Only the foresters working for the federal government could best oversee these forests to achieve the public good: "Since otherwise they will not do so, private owners of forest land must now be compelled to manage their properties in harmony with the public good. . . . The need for governmental control on private timberland is now self-evident. . . . The field is cleared for action and the lines are plainly drawn. He who is not for forestry is against it. The choice lies between the convenience of the lumbermen and the public good." The chief forester at the time, Henry Graves, was also a strong advocate of federal regulation: "The public in its own protection should prohibit destructive methods of cutting that injure the community and the public at large." Graves believed that the Forest Service "was an agency of sound policy in the public interest."[16]

Pinchot took the lead in defining the public interest in terms of conservation. In a letter he wrote to the secretary of agriculture (for the secretary's signature) concerning the management of the national forests once they had been transferred to the Agriculture Department, Pinchot wrote: "In the administration of the forest reserves it must be clearly borne in mind that all land is to be devoted to its most productive use for the permanent good of the whole people, . . . and where conflicting interests must be reconciled the question will always be decided from the standpoint of the greatest good of the greatest number in the long run."[17]

In a latter work, Pinchot discusses the three main principles of conservation: "development, preservation, the common good." The first principle is that conservation "stands for development. . . . Conservation demands the welfare of this generation first, and afterward the welfare of the generations to follow." He continues: "In the second place conservation stands for the prevention of waste." And finally, "The natural resources must be developed and preserved for the benefit of the many, and not merely for the profit of the few," what he later refers to as the "common good." Summarizing, he writes: "Conservation means the greatest good to the greatest number for the longest time." It is this summary statement, previously mentioned in the 1905 transfer letter, that is the basis of the utilitarian nature of this conception of the public interest. Throughout the history of the Forest Service this statement has often been cited by the organization's leadership and employees to justify their decisions.[18]

This utilitarianism meant that the forest lands should be managed, not simply preserved. At first, Pinchot and the technocratic utilitarians were united with the preservationists, both supporting the idea of public ownership. But they soon split, most pointedly on the proposal to dam Hetch Hetchy Valley in Yosemite National Park to supply water to San Francisco. The preservationists strongly opposed the project. Pinchot, however, favored it: "As to my attitude regarding the proposed use of Hetch Hetchy by the city of San Francisco . . . I am fully persuaded that . . . the injury . . . by substituting a lake for the present swampy floor of the valley . . . is altogether unimportant compared with the benefits to be derived from its use as a reservoir."[19]

The positions of Pinchot are echoed in the words of his friend and supporter Theodore Roosevelt. In a 1903 speech to the Society of American Foresters (SAF) at Pinchot's home, he elaborated on his forestry thinking: "Your attention must be directed not to the preservation of the forests as an end in itself, but as a means of preserving and increasing the prosperity of the nation. 'Forestry is the preservation of the forests by wise use.' . . . A forest which contributes nothing to the wealth, progress, or safety of the country is of no interest to the government and should be of little interest to the forester."[20]

The technocratic utilitarian perspective within the Forest Service continued following the Pinchot era. He left behind foresters who thought that "they were experts who alone knew how the nation's timber ought to be managed—'scientific' gentlemen who believed that their principles were wholly technical and free from self-interest." In the 1933 Forest Service report *A National Plan for American Forestry*, the Forest Service concludes that "practically all of the major problems of American forestry center [are] in, or have grown out of, private ownership. . . . A question raised very sharply by this breakdown in private ownership is whether the whole conception of transferring to private ownership so much forest land was not in error." In another 1933 publication, Robert Marshall of the Forest Service expresses further criticism of private ownership of the forests: "In the final analysis, public ownership of forests is absolutely necessary . . . [because] the nation is headed straight for forest bankruptcy under private ownership." Both of these passages suggest that the Forest Service continued to think that it was the government foresters, the technocrats, who were best suited to manage this land in the public interest.[21]

This commitment to technocratic utilitarianism continued into the 1960s. For example, in 1960, chief forester Richard McArdle cited the continued relevance of Pinchot's claim to manage the forests for "the greatest good to the greatest number for the longest time": "These instructions have constituted Forest Service doctrine from the beginning. They are the genesis of multiple use." The combination of multiple use (managing the lands for a variety of uses—fish and wildlife, forage, recreation, timber, and watersheds) and sustained yield, codified in the 1960 Multiple Use and Sustained Yield Act, formed the basis of technocratic utilitarianism following 1960. Indeed, "the myth of the omnipotent forester" continued in the 1960s. One forestry professor addressed his class as follows: "We must have enough guts to stand up and tell the public how their (sic) land should be managed. As professional foresters, we know what's best for the land."[22]

Although the forestry profession has backed away somewhat from its technocratic utilitarian perspective, trying to be more responsive to public demands, it continues to hold to the core of technocratic utilitarianism. For example, in 1976 the SAF distributed material to radio and television stations using "a variety of skillful ways . . . to explain why forest lands must be managed by professionals educated in the sciences of forestry." In 1988 testimony concerning management of a national forest in Oklahoma, the SAF representative was critical of prescriptive language that would limit the authority and flexibility of forest managers. "The Society is also concerned with the language that outlines specific timber management prescriptions for those areas that would be available for multiple use management," the testimony reads. "These prescriptions severely limit resource managers' abilities to take advantage of options that may improve the production of the forage and water, or the control of insects, disease, and fire, along with the myriad other multiple-uses."[23]

Since the establishment of forestry in the United States, the profession of forestry, forestry schools, and the Forest Service have been closely intertwined. The SAF, founded in 1900 by Fernow and Pinchot, among others, is the professional forestry organization. In 1898, the first forestry school was founded, and these schools have been closely linked with the SAF since 1900 (e.g., the SAF accredits forestry schools). Forest Service requirements largely determined instruction at these forestry schools. Henry Clepper reports that "a major influence on all schools was the written examinations given by the U.S. Civil Service Commission to recruit junior foresters for positions in the federal government. . . . A controlling, if unacknowledged, force in subject matter teaching was its contribution to the ability of students to pass examinations." Hence, both the SAF and the forestry schools have served to further the idea of technocratic utilitarianism, helping to create an internalized, shared value orientation within the Forest Service. The importance of this shared value orientation is confirmed by Herbert Kaufman in his study of the Forest Service behavior in the 1950s.[24]

Forestry was the dominant natural resources management profession advocating technocratic utilitarianism, but it was not the only one. Range management was greatly influenced by forestry; as it has sought to model itself after the forestry profession it also has been guided by the idea of technocratic utilitarianism. Additionally, the wildlife management and water management professions have been strongly influenced by this idea.

Although the technocratic utilitarianism perspective was at its peak in the early 1900s, the perspective has been institutionalized to a significant degree both within the forestry profession and within the land-management agencies of the federal government. Its place within the state is strong. Even though technocratic utilitarianism is no longer unchallenged in the natural resources management professions, it remains the core idea upon which these professions are based.[25]

Preservationism

The preservationist idea holds that certain lands should be retained by the government and preserved for moral reasons, their scenic beauty, and recreational purposes. It holds that private ownership, which preservationists associate with personal interest, profit motivation, and a short-term perspective, is incapable of the preservation of wild places. The beginnings of this preservationist idea regarding the public lands in the United States can be traced to the romantic movement of the eighteenth and nineteenth centuries. Prior to this time, nature was viewed as something to be conquered, something to be replaced with civilization. A significant theme of romanticism was that wilderness became viewed as the source of mystery and spiritual development, a place to more closely discover God and inner understanding.[26]

The European-based romantic movement influenced American thinkers and artists as well. In addition, an insecure America, lacking the artistic and historic treasures of Europe, pointed to the wilderness with a sense of pride as something in the United States that was grander and superior to anything found in Europe. As Roderick Nash writes, "In the middle decades of the nineteenth century wilderness was recognized as a cultural and moral resource and a basis for national self-esteem." He cites the writings of William Cullen Bryant and James Fenimore Cooper and the paintings of Thomas Cole, the Hudson River school, and western landscape artists as examples of the influence of nature on culture.[27]

The greatest midcentury spokesperson for the values of nature in the United States was Henry David Thoreau, who in 1851 proclaimed that "in Wildness is the preservation of the World." Thoreau reflected the thinking of a new intellectual movement in the United States (although one clearly owing much to the romantic movement), transcendentalism. This movement expressed discontent with progress, growth, and materialism. It turned to nature as a place where universal spiritual truths were reflected. For Thoreau, "wilderness was the source of vigor, inspiration, and strength. It was . . . the essential 'raw-material of life.'" In the end, wilderness was important "for its beneficial effect on thought."[28]

Around the time of Thoreau's writings, some began to call for the preservation of nature, which in their view was rapidly being destroyed. In the 1870s and 1880s, these proponents of preservation had their first successes. In 1872, Yellowstone was established as the nation's, and the world's, first national park. It was not established to preserve its wildness; rather, it was established to protect the scenic areas and natural wonders from falling into private hands, public interests that would be best served by government ownership. The state of New York acted in 1885 to establish a 715,000-acre forest preserve in the Adirondack Mountains, primarily for watershed protection, but also for scenic and recreational reasons. These lands were to be kept forever wild, a designation included in the state's constitution in the 1890s.[29]

A new spokesperson for preservation appeared on the scene in the last few decades on the nineteenth century: John Muir, a founder of the Sierra Club (in 1892). He was enthusiastic, and he became involved in politics to achieve his goals, which focused on government ownership as a means for preserving the Sierra Nevada and other wilderness areas. At first, Muir was sympathetic to active forest management, as advocated by Gifford Pinchot, which he viewed as an improvement over unregulated forestry. He soon came to realize, however, that forest management and preservation were not compatible. Muir, sensing that the forest reserves might be lost for preservation, focused most of his energy on the national parks. (The beginning of the dissolution came in 1897, when Muir and Pinchot clashed over grazing on the forest reserves.) Supporters of preservation followed suit, focusing on creating and protecting national parks, and later, wilderness areas. Public support for preservation continued to grow during this period, primarily in response to the closing of the frontier, industrialization, and urban problems. In addition, wilderness served as a source of esthetic and ethical values.[30]

The real split between the supporters of technocratic utilitarianism and preservationism came in the early twentieth century during the debate over Hetch Hetchy Valley within Yosemite National Park. The city of San Francisco requested permission to dam the Tuolumne River and flood the valley to use as a source of water. In 1908, Secretary of the Interior James Garfield approved the request, setting the stage for congressional action. A spirited national debate followed, stretching to final congressional approval in 1913. Muir and the preservationists fiercely fought this proposal because it was an attack on a national park, an area set aside to be preserved, stressing the spiritual and scenic aspects of the valley. Muir claimed, "These temple destroyers, devotees of ravaging commercialism, seem to have a perfect contempt for Nature, and instead of lifting their eyes to the

God of the Mountains, lift them to the Almighty dollar." Pinchot and the supporters of the reservoir responded that human use of natural resources, even in parks, must take precedence. Despite the flooding of Hetch Hetchy, preservation had received a national airing and had fared rather well. Preservationists aroused public opinion, began to flex their political muscles, and made preservation a national issue. Just three years later, in 1916, the National Park Service was created to oversee the expanding national park system.[31]

In the 1920s, preservationists began to make inroads on the national forests. In 1920, landscape architect Arthur Carhart recommended that the Forest Service manage Trappers Lake in Colorado as wilderness, rather than develop vacation homes on the site, a recommendation that was followed. Forester Aldo Leopold recommended that a large area in the Gila National Forest in New Mexico be preserved as wilderness, which he defined as "a continuous stretch of country preserved in its natural state, open to lawful hunting and fishing, big enough to absorb a two weeks' pack trip, and kept devoid of roads, artificial trails, cottages, or other works of man." A 574,000-acre area was established there in 1924. During his life, Leopold expanded preservationist thinking, adding an ecological, scientific basis for preserving the land (to accompany the existing sentimental and spiritual bases). He also advocated the development of a land ethic, of which preservation was a part. And in 1926, the Forest Service established a wilderness area in the lake and forest country of northern Minnesota. Furthermore, the Forest Service adopted a set of regulations (the L-20 Regulations) governing wilderness throughout the national forest system in 1929.[32]

The commitment to preservation continued to grow in the 1930s, as the supporters of preservationism continued to press for the establishment of more wilderness areas. Robert Marshall was a national leader in this effort. As head of the Division of Recreation and Lands in the Forest Service, he helped draft the U-Regulations, which replaced the L-20 Regulations, and helped to preserve fourteen million acres of forest lands. He was also a strong supporter of public ownership, advocating federal ownership of almost all of the forest lands then in private hands: "If private ownership leads to anti-social conditions, and brings about undesirable results, there is no more reason why it shouldn't go by the boards than slavery went by the boards." In 1935, Marshall, Leopold, and others formed the Wilderness Society, the purpose of which was "to integrate the growing sentiment which we believe exists in this country for holding wild areas sound-proof as well as sight-proof from our increasingly mechanized life. . . . [These areas are a] serious human need rather than a luxury and plaything." This organization has been in the leadership role in the fight to preserve public lands ever since.[33]

The supporters of the idea of preservation continued to achieve success in the political world. In the 1950s, preservation forces achieved their greatest victory in blocking the proposed Echo Park dam, which would have flooded part of Dinosaur National Monument in Colorado and Utah. This victory was in sharp contrast to the flooding of Hetch Hetchy Valley in the 1920s. Building on the success at Echo Park, the preservation movement focused its energies on passing a law to give statutory protection to the preservation of wild lands. The first bill was introduced in 1956, and after much debate and compromise, a wilderness law was passed in 1964. The preservationists relied on strong grass-roots support to gain success against the opposition of wood-using industries, oil and gas interests, grazers, mining interests, and most professional foresters.[34]

Preservationists could not rest on this victory, however. In the 1960s, the government proposed two dams in the Grand Canyon as part of a comprehensive water plan for the Southwest. Preservationists launched an all-out attack on the dams. The Sierra Club ran full-page advertisements in major papers opposing the dams (e.g., "Now only you can save the Grand Canyon from being flooded . . . for profit" and "Should we also flood the Sistine Chapel so tourists can get nearer the ceiling?"). Howard Zahniser of the Wilderness Society summarized the position of the preservationists: "Out of the wilderness has come the substance of our culture, and with a living wilderness . . . we shall have also a vibrant, vital culture, an enduring civilization of healthful, happy people who . . . perpetually renew themselves in contact with the earth. . . . We are not fighting progress, we are making it." Once again, the preservationists were successful and the proposed Grand Canyon dams were deleted from the overall project. In 1968 the Wild and Scenic Rivers Act was passed, establishing a framework for protecting wild rivers.[35]

The next major battle came in the 1970s over, in the minds of preservationists at least, the biggest prize: Alaska. As a result of the Alaska Native Claims Settlement Act of 1971, Congress was to determine which lands should be set aside as "national interest" lands, and hence be unavailable to the state of Alaska and natives in their land selection processes. When Congress failed to act by the 1978 deadline, President Jimmy Carter directed the withdrawal of 110 million acres from mineral entry and state selection and of 11 million acres of national forest land from mineral entry, gave national monument status to 56 million acres of land under the Antiquities Act, and created 40 million acres of new wildlife refuges. The Wilderness Society called it "the strongest and most daring conservation action by any president in American history." The Carter reservations were designed to be temporary, until Congress acted. Preservationists were well organized for this fight: five major environmental groups—the Wilderness Society, the Sierra Club, Friends of the Earth, the National Parks and Conservation Association, and the National Audubon Society—formed the Alaska Coalition. Before the battle was over, the coalition included more than fifteen hundred groups with a combined membership of ten million. As Nash writes, "Wilderness never had more friends" than in this quest. In the debate over how much of Alaska to preserve, Representative John Seiberling of Ohio referred to Alaska as "part of the heritage of mankind," and John Kauffman of the National Park Service said that "Alaska is our ultimate wilderness, the last remnant of what the New World used to be. If we lose the freshness and the beauty there, something essential to North America will have died out forever." Preservationists were reasonably successful in the Alaska fight as well: 104 million acres (28 percent of the state) were set aside for permanent protection, and 56.7 million acres were immediately given wilderness status. An additional 70 million acres of federal land were to be reviewed for future wilderness designation. It was the largest single preservation action in the history of the world.[36]

What, then, is the justification for preservation used by the movement? After this brief history, some of the justifications are apparent. It should also be apparent that the justifications have changed, typically expanding over time. There are nine major rationales. First, the transcendentalist ideas, which can be traced back to the 1800s. As Wilderness Society cofounder Benton MacKaye claimed, preserved nature is a place "to recharge depleted human batteries directly from Mother Earth." Sigurd Olson wrote that "only through my own personal contact with civilization had I learned to value the advantages of solitude." Thus nature is a place for spiritual recharge and is needed to allow us to better understand ourselves and civilization. A second, related rationale is that wild nature is a source for creativity. It can be argued that "the springs of artistic and intellectual creativity depend on fresh penetrations to and interpretations of this elemental reality [wilderness]." A slightly different twist to this is a third rationale: wilderness serves as a place to keep the human spirit alive, to resist the potential totalitarian societies of the future.[37]

Fourth, preserved nature is important historically to America; it is an important force on American thought and culture and should be preserved as part of what Wallace Stegner calls "the geography of hope." More recently, as people have begun to more clearly see their relationship to nature and the environment, and have even begun to fear for their survival, some have looked to nature as an example. A fifth rationale asserts that nature serves to develop a sense of humility and restraint in human beings. A sixth rationale, related to the newfound concern with our environment, is that preserved nature can serve as a reservoir for healthy ecological functions and biological diversity.[38]

Seventh, a growing literature cites the benefits of wilderness for mental health, especially as a treatment method. An eighth rationale for preserving nature is that because it is scarcer, it is naturally more valuable. That is, simple scarcity theory suggests that as something becomes less common, it will have a higher value for society and society will make stronger efforts to preserve it. The final major rationale is relatively new: the right of nature to exist for its own sake, not because humans deem it worthy of protection. I think that this rationale is fundamentally different from previous rationales, because it is not based on preserved nature's value for humans.[39]

Why do the preservationists favor government ownership rather than private ownership? In the beginning, the preservationists, like the technocratic utilitarians, saw the free market as incapable of proper natural resource management and of the preservation of certain lands. Hence, they united with the technocratic utilitarians to keep lands in public ownership. Through the 1980s, preservationists continued to distrust the private sector for preservation and general management of the resources on the public lands. The following comments are representative of the preservationist position:

> We believe in the public lands. We strongly believe that they should remain public and should be maintained and managed for the public. (Friends of the Earth)

> American history has demonstrated that the public is not well served, in the long run, by turning over commodity lands to private interests. (Wilderness Society)

> We think of it [public lands] as a national legacy that should be owned by no one and enjoyed by everyone. . . . We want to see it protected and managed as public lands, by public agencies and available to the public. (Wilderness Society)

More specifically, preservationists favor government ownership primarily because they oppose private ownership of wild places, which they associate with personal interest, profit motivation, and a short-term perspective.[40]

The preservationist movement remains strong today. Among the major preservation-oriented groups, the Wilderness Society has 310,000 members, the Sierra Club has 650,000 members, the National Audubon Society has 600,000 members, and the National Parks and Conservation Association has 300,000 members (all figures for 1994), each figure an all-time high. Since the passage of the Wilderness Act, supporters of the preservation idea—many who were members of these groups—have worked to prevent dams, expand the wilderness system, create new parks, establish a wild and scenic rivers system, and preserve Alaska.

The idea of preservation has been influential since the late 1800s. Beginning in the 1960s, however, and continuing through the 1990s, it has drawn increasing numbers of supporters and become even more powerful politically (as victories on wilderness and Alaska attest). There is nothing to suggest that this idea will soon decrease in popularity; rather, it seems that it will persist, or even grow, in importance.

Conclusion

From the late 1800s through the present, these three ideas have been dominant in the public-lands policy regime. As I will demonstrate in the cases to follow, these ideas have served to shape and constrain policy debates, helping to give each policy regime a particular pattern. In each policy regime, one of these ideas was embedded within the state and has become privileged regarding the other ideas in policy debates. The ideas that were not embedded within the state did not wither away, however. Supporters of the other ideas have launched challenges to the privileged ideas, and, beginning in the 1960s, the foundations of privilege began to erode. I now turn to a series of case studies in hard-rock mining, forestry, and grazing policy that demonstrate this dynamic at work.[41]

MINING AND
THE LAND
MINERALS AS
THE ULTIMATE
RESOURCE

This chapter is the first of three examinations of natural-resources policy on the public lands. In this examination, as in the two following chapters, I present three particular case studies. The first case is an examination of the formation of the hard-rock mining policy regime, the period from 1848 to 1872. It was during this period that the state was first organized within the policy regime, that a particular idea became embedded within the regime, and that the foundations for the privatized policy pattern were laid. The second and third cases are examinations of significant challenges to the institutionalized idea and the existing policy pattern. In both the passage of the Wilderness Act of 1964 and the strategic minerals debate from 1975 to 1985, supporters of preservationism led challenges to the privatized mining policy. In both cases, inroads were made by preservationism, and the privileged position of economic liberalism within the mining policy regime was weakened. Nevertheless, it remains the privileged idea. Through these cases, covering 125 years, we can see how an idea becomes embedded within a policy regime, how that idea serves to shape the policy pattern in that regime, and how difficult it is to dislodge that idea.

The Mining Laws of 1866, 1870, and 1872: Laying the Foundations for Mining Policy

The discovery of gold in California in 1848 ushered in a series of mineral rushes throughout the American West. At that time the federal government had no laws dealing with mining on the federal lands, despite the large scale and scope of such mining. The federal policy was one of laissez-faire until a series of laws were passed, beginning in 1866. The first law dealt with quartz or lode mining, an 1870 law dealt with placer mining, and the 1872 general mining law reaffirmed the earlier two laws and built upon them. The 1872 Mining Law continues to be the legal framework under which hard-rock minerals are exploited on the public lands.[1]

When gold was discovered, California was under the control of the U.S. Army. The state was still technically a part of Mexico, although United States troops had occupied it beginning in 1846, during the Mexican-American War. Col. R. B. Mason, military governor of the territory, did not attempt to remove the miners as trespassers on federal lands due to more pressing concerns and a lack of personnel. He reported that "upon considering the large extent of the country, the character of the people engaged, and the small scattered force at my command, I am resolved not to interfere, but to permit all to work freely, unless boils and crimes should call for interference."[2]

Mexican law remained in force during the occupation, until February 1848, when Colonel Mason proclaimed that "from and after this date, the Mexican laws and customs now prevailing in California relative to [acquisition of mining rights on public lands] are hereby abolished." As part of the treaty ending the Mexican-American War, lands in the present southwestern United States were transferred to the United States government. No alternative legal system was offered in place of the Mexican laws, and it was not until September 1850 (when California was admitted as a state) that the laws of the United States became officially binding in California. Even then there was no mechanism allowing for the lawful exploitation of minerals from the public lands. According to federal law the miners were trespassing on public lands.[3]

In addition to this lack of law, the military forces charged with keeping the law in California were minimal. These forces would not have been able to enforce mining laws even if they existed, especially once the population of California swelled with the arrival of tens of thousands of fortune seekers over the next few years. The minimal military force was further reduced, to about six hundred men, due to desertions by soldiers in search of gold. Hence, even if the military were disposed to enforce trespassing laws, it did not have the resources to do so, and there is no evidence that the military interfered with miners from 1848 through 1866.[4]

Miners were unwilling to wait for Congress to act, but they did need some degree of order to proceed with the mining. The miners acted to fill the legal vacuum by creating mining districts and mining codes to govern actions within these districts; that is, they established an extralegal system to govern mining activities on these public lands. The early agreements among the miners were sharing contracts; miners did not receive exclusive control of any particular piece of land, rather groups of miners shared the output of all miners in the group. These agreements soon changed, primarily due to the increased population in the mining areas.[5]

The new agreements were based on specific, exclusive land allotments to miners; the gold now went with the land a miner controlled. The focal point of these mining codes was a method of granting property rights to miners on this land that actually belonged to the federal government. The miners wanted to make sure that whomever discovered the gold and worked a particular area received some right to mine that claim without having to worry that someone else would claim the same area, or try to force him off the area with violence.

These mining codes, although distinctive from one locality to the next, did share a number of general characteristics. Among the most common were: (1) the definition of the geographic boundaries of the district; (2) the stipulation of the maximum size of the claim (usually seeking to limit the size to prevent monopoly, based on the "common aversion of frontier democracy to monopoly"); (3) the appointment of a leader and claim recorder; (4) the adoption of a code of laws; (5) an agreed way to mark claims and determine the consequences of trespassing; (6) the assignment of exclusive property rights to a claim to an individual; (7) the statement of conditions that must be met to maintain an exclusive claim (e.g., amount of work per week) and the circumstances by which rights to a claim would be forfeited; (8) the indication of the maximum number of claims an individual could have; and (9) the establishment of a means to settle disputes and enforce rules and decisions. Despite the crudeness and ephemeral nature of some of these districts and codes, the property rights granted in the codes were widely respected.[6]

As the miners developed these mining codes, the federal government debated what federal policy on these lands should be. Following reports from Colonel Mason, President James Polk recommended that the mineral lands either be kept for the government's use or sold in small quantities at a fixed price to generate income for the federal treasury. The Senate Committee on Public Lands reported a bill that would have allowed for the purchase of two-acre plots at not less than $1.25 per acre. This was opposed by Senator Thomas Benton (D, Mo.), a chief supporter of mining interests. He opposed any effort to raise money from the mines and instead proposed a nonfee permit system. Neither bill passed. President Zachary Taylor, like his predecessor, favored the sale of these lands. In a December 1849 message to Congress, he advocated dividing the mineral lands into small tracts for sale or lease, but again Congress did not act.[7]

Congress returned to the mining issue in 1850. Newly seated Senator John Frémont of California (D) introduced a bill based on Senator Benton's plan. Federal agents would oversee the mining, with permits being granted to American citizens who wanted to mine. Though it charged miners no fee, the bill was strongly opposed by the California miners, who resisted any federal interference. In debate on the bill, Senator Thomas Ewing (W, Ohio) argued that the federal government should raise revenue from the lands and proposed that miners be required to sell their gold to the government, which would then sell it at a profit. An amendment to this effect was opposed by Benton and the California senators and was defeated. Senator Alpheous Felch (D, Mich.) was opposed to any kind of leasing or permitting system, pointing to the failures of the earlier lead-leasing program. He argued for the status quo: continued freedom to the miners with no government interference. The majority of senators favored some degree of regulation, at least to maintain order. Eventually, Frémont's bill passed. It died, however, when no legislation emerged from the House.[8]

Millard Fillmore became the third president to recommend that Congress dispose of the mineral lands in the West. He recommended that the lands be divided into small tracts and sold to benefit the treasury. He also advocated that precautions should be taken to prevent concentration of landownership. This proposal, like the others, was strongly criticized in California. Opponents found the policy undemocratic and favorable to larger interests, leading to the potential of monopoly. The California legislature passed a resolution declaring that "the policy of selling the mineral lands would be in conflict with the true interests of the state and nation, for the richest mineral lands would fall into the hands of speculators, resulting in the stoppage of immigration and the retardation of the progress of California." Californians also warned the federal government that the miners "had become accustomed to consider the mineral lands as a common heritage, and would not brook any interference." The state legislature advocated either the continued status quo or turning the mineral lands over to the state for regulation.[9]

After considering this strong opposition, President Fillmore abandoned his former recommendations and in December 1851 recommended that the gold mines "be permitted to remain as at present, a common field, open to the enterprise and industry of all our citizens, until further experience shall have developed the best policy to be ultimately adopted in regard to them. . . . It is safer to suffer the inconvenience that now exists, for a short period, than by premature legislation to fasten on the country a system founded in error, which may place the whole subject beyond the future control of Congress." Hence, the federal government continued its support of the laissez-faire approach. The issue of federal mineral lands policy was dormant until 1858.[10]

During the debate over mineral policy in the period from 1848 to 1872, the dominant idea regarding public-lands management was economic liberalism. This idea was virtually unquestioned. The other two ideas, technocratic utilitarianism and preservationism, had yet to make themselves felt. The technocratic utilitarian perspective, in the guise of forestry, was flourishing in Germany and other parts of Europe at this time, but it had no significant effect in the United States. Preservationism, meanwhile, was in its infancy. So the debate focused on how best to get these resources into private hands (while perhaps gaining some money for the federal treasury).

The general economic liberalism of the early and middle 1800s was not strictly laissez-faire, not a time for "the jealous limitation of the power of the state," in the words of James Willard Hurst, but rather "the release of individual creative energy was the dominant value." This desire for the release of energy was especially pronounced in the case of natural resources, such as minerals. The dominant idea in mining, reflecting this mode of thinking about the relationship of state and society at the time, was that society generally would be best served by government action to stimulate the finding and mining of minerals, which would in turn stimulate economic growth and western settlement. This system served as an incentive to stimulate prospecting and the subsequent discovery and production of minerals, which in turn fueled the growth and development of the West and of the national economy. The laissez-faire argument had held that only if a prospector knew that he would receive the profits from his discovery would he expend the necessary labor and capital to discover the minerals. If this incentive were removed, the discovery of minerals, and economic growth, would slow. As the miners themselves were creating extralegal systems to govern their mining districts, the federal government, this argument ran, should act only to legitimize and help enforce these mining codes when necessary. It would, in other words, aid in "the release of energy." Any further government involvement would lead to chaos and confusion and would not serve the public interest, as was demonstrated by the ill-fated effort to lease lead lands in the 1820s and 1830s.[11]

The mining activity in Nevada soon rivaled that of its neighbor in importance, both in terms of production and politics. Nevada mines, especially the Comstock Lode, were capital-intensive, and investors demanded secure property rights, an issue complicated by the complex pattern of underground veins. Mineral rushes continued throughout the rest of the western United States as well, and by 1866 there were more than eleven hundred mining districts in Arizona, California, Idaho, Montana, Nevada, New Mexico, Oregon, and Utah.[12]

In 1858 the federal government returned to the issue of mining policy. Interior Secretary Jacob Thompson recommended that Congress adopt some definite policy regarding the lands. California was opposed to any such congressional action, with one editorial going so far as to claim that federal government interference in mining "would result in the loss of California to the Federal Union." In response to the interior secretary's recommendation, California senator William Gwin (D) introduced an amendment to the homestead bill to legalize the current situation. The amendment was rejected due to concerns that it was a virtual cession of mineral lands to California with no financial gain for the federal government and due to concerns of germaneness.[13]

The mining issue took on newfound importance during the Civil War years as the federal government sought new sources of revenue to help pay for the war. Four major proposals were explored in Congress: taxation, sale of the lands, governmental regulation, and governmental control. In 1864, the House passed a bill to tax the production of valuable metals 5 percent. The Senate debated the issue extensively, and, predictably, mining interests and the western senators were strongly opposed to the tax. The Senate passed a bill with .5 percent tax on proceeds, which the House agreed to. The tax became law, but it never raised much revenue and was repealed in 1866. The sale of the lands, in small parcels, was championed by Representative George Julian (R, Ind.), chair of the Committee on Public Lands. He introduced a bill in early 1865 that called for dividing the lands up into small tracts and selling them at auction, with a set minimum price. Miners who already had claims would be given first chance to purchase their claims at these minimum prices. The bill was designed both to raise revenue and encourage orderly settlement of the public domain. Due to strong opposition from California and Nevada legislators, neither Julian's bill nor a companion bill introduced by Senator John Sherman (R, Ohio) received serious consideration.[14]

The proposal for government regulation came from Representative John Kasson (R, Iowa) in late 1865. This system would have allowed miners to continue to mine on federal lands, but they would have had to file their mining returns with the federal government and would then pay an established income tax on this amount. The bill died in committee. The final proposal called for the president to take over the gold and silver mines and use the proceeds to pay for the war and war debts. This proposal did not go very far for numerous reasons. It must be emphasized that no one seriously advocated a leasing system at the time, recalling the failures of the lead-leasing program, referred to by Representative Julian as "the experiment [that] failed utterly." Treasury Secretary Hugh McCulloch denounced leasing the mines as "impracticable, un-American, and unconstitutional."[15]

Despite the failure of any of these proposals to be adopted, a consensus had developed that the government had to adopt some policy regarding the mineral lands; the laissez-faire approach could no longer continue. To this end, both the Senate and House created committees on mines and mining in 1865. This marked the termination of the first approach of economic liberalism to mining, a wholly laissez-faire approach.

A second phase of economic liberalism, less laissez-faire, accompanied the increased importance of the more capital-intensive quartz mining. In order to make such mining endeavors more attractive to investors, the mining interests wanted to make the titles to these mines more secure. They wanted the federal government to go beyond the existing approach: some system should be established for the acquisition of legal titles to mineral lands. Investors were skeptical of the ephemeral mining districts and codes, even with government acquiescence. The push for federal mining legislation came from these capital-intensive quartz mining interests, not from the more numerous, independent placer miners. The 1866 mining law, dealing only with quartz mining, provides evidence that this was the case.

During the debates of 1865 and 1866 on the different mining-lands bills, the western miners and states made it clear that they were opposed to a sale program. They favored a program that would cause as little disruption to mining as possible yet strengthen property-rights claims. The public interest, they argued, would be best served by transferring these federal mineral lands into the private sector at minimal cost and in compliance with existing mining codes. In April 1866, Senator Sherman (formerly an advocate of a sales program) introduced S. 257, which was assigned to the new Committee on Mines and Mining, chaired by Senator John Conness of California (R). Conness and the newly seated senator from Nevada, William Stewart (R), altered the bill significantly in committee. It emerged as a bill dealing with quartz mining that permitted continued free access and exploitation but allowed for the purchase of mineral lands for a minimal fee. The bill would also recognize the claims of those already on the lands. In reporting the bill out of committee in May, Senator Conness claimed that "the bill proposed adopts the rules and regulations of miners. . . . [They] are well understood, and form the basis of the present admirable system. Popular sovereignty is here displayed in one of its grandest aspects, and simply invites us not to destroy, but to put upon it the stamp of national power and unquestioned authority." Full Senate debate on the bill began in June. Senator Stewart, who had become the chief advocate for mining interests, and other supporters stressed the importance of security for quartz mining claims, in order to attract needed capital investment. Senator Stewart echoed Senator Conness: "All there is in this bill is a simple confirmation of the existing conditions of things in the mining regions, leaving everything where it was, endorsing the mining rules." Senator George Williams (R, Ore.) introduced an amendment seeking to maintain the status quo. He claimed that the Stewart bill would favor the large miners: "Nine-tenths of the men who are today engaged in mining are opposed to this bill or any other bill of a like nature contemplating a sale of these mines." The amendment was defeated. After further debate, the bill passed the Senate with the support of all the members from the mining committee.[16]

In the House, Representative Julian succeeded in having the bill assigned to his Public Lands Committee, where he let it linger because he favored the sales alternative. The senators from California and Nevada, led by Senator Stewart, then resorted to some legislative chicanery. They decided to use a bill that had already passed the House and was now under Senate consideration as their Trojan horse. The bill, H.R. 365, titled an "Act Granting the Right of Way to Ditch and Canal Owners over the Public Lands, and for Other Purposes," had come out of the House Committee on Mines and Mining. The western senators successfully moved to strike the text of the bill and substituted the text of S. 257 in its place. All that remained of H.R.

365 was its title. The bill passed the Senate, and when it went back to the House for reconsideration, it was assigned to the Mines and Mining Committee. Recognizing the trickery, Representative Julian attempted to have the bill referred to his Public Lands Committee, but failed. He strongly tried to block passage of the bill on the floor, but the western mining interests were able to generate enough support to pass the bill seventy-three to thirty-seven. (It is reported that Senator Stewart visited each member of the House to try to gain support for his proposal.) The first general mining bill became law on July 26, 1866.[17]

The opening section of the law reads: "The mineral lands of the public domain, both surveyed and unsurveyed, are hereby declared to be free and open to exploration and occupation by all citizens of the United States, and those who have declared their intentions to become citizens, subject to such regulations as may be prescribed by law, and subject also to the local customs or rules of miners in the several mining districts, so far as the same may not be in conflict with the laws of the United States." This section established the general framework of federal mining law at the time and through the present: an acceptance of existing customs and the continuation of free and open access to the public lands. The law allowed miners to file quartz claims in local land offices and to purchase this land (i.e., to receive a patent) for $5.00 per acre, providing that they had followed local mining customs and spent $1,000 in labor and improvements on developing the land. The key aspect of the law was the acceptance of the local codes and customs. The law legitimized and adopted the extralegal property-rights system developed by the miners. At last, property rights could be legally recognized and obtained from the federal government.[18]

In the next six years, Congress passed two additional mining laws. The 1870 law amended the 1866 act to include placer claims, allowing a person or group to patent up to 160 acres and to purchase these claims for $2.50 an acre. The 1872 Mining Law primarily recodified the provisions of the 1866 and 1870 acts. Although the interests of both the placer miners and quartz miners were served by these laws, the passage of the acts by a Congress in which the western states had limited power indicates that the idea reflected in these laws, economic liberalism, was one that dominated society at the time.[19]

With the passage of the 1872 Mining Law, the idea of economic liberalism became institutionalized and the hard-rock mining policy pattern of privatization had begun. In two important ways, such an outcome might have been inevitable. Importantly in this case, there was no significant competing idea of how the public interest would be best served in public-lands management. As noted above, neither technocratic utilitarianism nor preservationism existed as a developed idea of the public interest. Additionally, this idea of economic liberalism for the public lands was indicative of the general thinking about how state and society should interact. Although the centrality of "the release of energy" view was beginning to wane, it held on longer in the West, where settlement and commerce were just becoming established. And second, a national administrative state was virtually nonexistent in the 1860s and 1870s. The federal government did not have the capacity, even if it had the desire, to implement a more involved regulatory mining policy.[20]

Early mining policy, like most early natural-resources policy, was administered by the military. The lead-leasing program of the 1820s and 1830s was administered by the army, and the army was the only potential federal administrative actor in California and the rest of the West during the mineral rushes from 1848 through the 1860s. The military was charged with administering these programs not because of its expertise, but because of the lack of alternatives. Its lack of expertise, along with its numerous other responsibilities, made the army an unlikely viable mining management agency. Following the California gold rush, the army chose to allow the miners to regulate themselves rather than step in and remove them for trespassing or establish a federal system of law and order. The federal government had no other choice for administering mineral policy. The two oldest minerals-related agencies, the United States Geological Survey (USGS) (1879) and the Bureau of Mines (1910), did not yet exist. The General Land Office (GLO) in the Department of the Interior was a land-disposal agency, staffed almost wholly by clerks. Also, the GLO did not have offices in much of the West because it established new offices gradually, as settlers demanded agricultural land.[21]

The existence of the extralegal property-rights system developed by the miners in the various mining codes they established helped to pave the way for the privatization of hard-rock mining policy. These extralegal property rights demonstrated an existing order within the private sphere, an order the federal government did not want to undermine. Hence, the federal laws accepted these existing property-rights schemes and incorporated them into federal law. Even though the policy was essentially privatized, the state still played a major role as the guarantor of these property rights.[22]

The passage of the 1866, 1870, and 1872 hard-rock mining laws reflected the dominance of the economic liberalism conception of the public interest, the nonexistence of a national administrative state, and the existence and subsequent adoption of an extralegal property-rights system. Overall, the 1872 Mining Law has continued to be the centerpiece of the privatized policy regime for hard-rock mining policy on the public lands. This fact is testimony to the institutionalization of economic liberalism, despite the rise of preservationism, a massive administrative state, and changes in the public's conception of property rights.

Mining and the Wilderness Act: Institutionalized Power and the Special Treatment of Mining

The first significant challenge to the privatized policy regime of hard-rock mining policy came in the late 1950s and early 1960s in the battle for passage of the Wilderness Act. Supporters of preservationism sought to have large areas declared wilderness and made off limits to development, including mining. Though not a direct assault of the privatized policy pattern, it threatened the privilege of economic liberalism. Mining interests strongly opposed a wilderness law, and when a law passed in 1964, minerals were the only commodity that could still be developed.

There had been a few changes in the policy regime since the 1870s, but the privilege of economic liberalism remained institutionalized. Two institutions were created in the forty years following the passage of the Mining Law of 1872, both of which held the potential to give the state the capacity to retreat from the privatization of mining policy. The idea of economic liberalism, however, had already become embedded within the state, and the USGS and Bureau of Mines supported this existing system. The USGS focused on geologic mapping and helping to locate mineral deposits for the mining industry. George Otis Smith, who became director in 1907, said that "in the Survey's efforts to serve the mining industry, then, I recognize no limitations." Smith also said that the first director of the USGS, Clarence King, "regarded the bureau as being charged with the duty of directly serving the mining industry of the country." A review of the history of the USGS finds these statements to ring true. It should be stressed, though, that neither director saw a conflict in serving the mining industry and the public interest. By helping the industry, they were helping the country. The Bureau of Mines was established in 1910 in response to requests made by the metal and coal mining industries. Its mission focused on mine safety, mine technology, and the economic conditions of mining, especially on the public lands.[23]

The continued free access of mining interests to wilderness, wild, canoe, and primitive areas designated by the Forest Service on the national forests was one of the main reasons that preservationists sought to establish a statutory wilderness system. The Forest Service could prevent most development within these areas, but these lands, like all national forest lands, remained open to mining activity under the Mining Law of 1872. Although these areas were not typically mined, it was not due to Forest Service controls, but to the remoteness or inaccessibility of the locations.[24]

As originally drafted by Howard Zahniser of the Wilderness Society in 1956, mining and prospecting in wilderness areas were to be prohibited except on claims already established. In a later draft, he added language that would allow the president to open wilderness areas for mining if necessary for national security. With this language, Zahniser hoped to reduce opposition from mining interests. In his final draft, however, this presidential latitude was eliminated and the original language was reinserted. Hence, even before the bill was introduced in Congress, wilderness supporters were well aware that mining interests would be among the chief opponents to the bill.[25]

The first hearings on a wilderness bill were held in the Senate in June 1957, focusing on S. 1176, introduced by Senator Hubert Humphrey (D, Minn.). The bill stated that "no portion of any area constituting a unit of the National Wilderness Preservation System shall be devoted to commodity production . . . prospecting, mining or the removal of mineral deposits (including oil and gas)." The preservationists backing wilderness thus began with a bill that did not contain any compromises. As the legislative process continued, however, the problem of mining appeared to be one of the chief obstacles to the passage of a bill, so compromises were made.[26]

The American Mining Congress (AMC), the chief group representing the mining industry, testified against the bill, arguing that it hurt the mining industry in particular and the nation in general. During this first testimony, the mining industry developed a number of themes of opposition that were to become standard throughout the debate over wilderness. The first theme was that the wilderness bill, like any other withdrawal of public lands, ran counter to the historical free access of public lands by mining interests (the privatized policy regime). Because "minerals are where you find them," as much land as possible should remain open for exploration. The wilderness bill was part of a continuing trend of reducing the amount of land open to such exploration and development.

Mining representatives also argued that such a law was not needed. Plenty of wilderness existed and always would exist because of the ruggedness of the terrain in the West. In addition, the Forest Service had an administrative wilderness program in place that was functioning well. A third criticism of the wilderness bill was that it contradicted the policy of multiple use. Large areas of land would be set aside for one use: wilderness. The goal of public-lands management should be multiple use. A fourth theme of the mining opposition was that these lands should remain open for scientific reasons; declaring them wilderness might prevent unforeseen scientific advances.

The two final themes of opposition were the most important: economic development and national security. The mining community argued that reducing the land available for exploitation would slow the development of minerals at a time of economic and population growth. This would slow the United States economy and make the nation even more dependent on imported minerals. This point ties in with the national security concerns. A reliance on imported minerals could be catastrophic during a time of war. The United States must develop domestic sources for minerals to the fullest extent possible. The nation did not have the luxury to close off areas from development, especially when we did not know if any minerals existed in them. The AMC testimony concludes: "We in the mining industry are unalterably opposed to the locking up of natural resources of any kind from development for the public good. . . . We urge you to disapprove the measures now pending before you which would establish a public system devoted to national wilderness preservation and would prevent development of those areas now open for mineral location."[27]

In general then, this debate was between the supporters of preservationism and the supporters of economic liberalism, who had their idea institutionalized within the state. The passage of the wilderness bill was the primary political objective for the supporters of preservationism because it would statutorily preserve large areas of land that they thought had special significance. The AMC testimony demonstrates the position of supporters of economic liberalism. Wilderness would prevent free access to minerals on the public land, preventing the free market from working, hindering economic growth, and potentially wasting valuable minerals. Clearly, such a policy was not in the public interest.[28]

A further institutional advantage for mining interests is reflected in property rights. Even prior to the passage of the Mining Law of 1872, mining interests proclaimed their property rights to minerals found on the public lands. As discussed in the prior case study, this extralegal property-rights system was legitimized, and mining interests now proclaimed their legal rights to these minerals. The wilderness bills were attacked because they would limit the ability of miners to discover and develop these minerals, a violation of property rights as granted under the 1872 statute.

In 1958, the Senate considered a new wilderness bill (S. 4028) that included language preservationists hoped would reduce the opposition of commodity groups to wilderness. Although still prohibiting commercial activity in general, the bill allowed for presidential exceptions: "Within national forest areas included in the Wilderness System the President may, within a specific area and in accordance with such regulations as he may deem desirable, authorize prospecting, mining, the establishment or maintenance of reservoirs and water-conservation works, and such measures as may be found necessary in control of insects and diseases, including the road construction found essential to such mining and reservoir construction, upon his determination that such use in the specific area will better serve the interests of the United States and the people thereof than will its denial." Despite this change in language, the AMC again testified in opposition to the bill, concluding with the same language it used in 1957: "We in the mining industry are unalterably opposed to the locking up of natural resources of any kind from development for the public good. . . . We urge you to disapprove S. 4028."[29]

In the autumn of 1958 and the spring of 1959, the Senate held a series of six hearings throughout the West to receive local input on the wilderness bill. Numerous local, state, and regional mining concerns testified at these hearings, all opposing specific wilderness bills and the general idea of a wilderness law. The groups that testified used arguments similar to those made by the AMC, especially focusing on locking up resources and the threat to national security. The Alaskan commissioner of mines asked, "What good are these resources if they must remain in their natural state? . . . Alaskans wish to manage their lands and develop their resources in the public interest. This can only be accomplished under the principles of multiple purpose use." The Northwest Mining Association claimed, "Our, yours, and my natural resources are being wasted by conservationists. They call this waste 'wilderness areas.' "[30]

This field testimony also indicated some new themes of mineral industry opposition. The presidential exemption was unacceptable because it would lead to little or no exploration in fact. There would be no incentive for such exploration because there was no guarantee that the finder of the minerals would be allowed to develop them (counter to the incentives of economic liberalism). Many groups, at these and future hearings, called this presidential exemption "meaningless" to mining interests. Two additional themes suggested that the mining interests were beginning to think that a wilderness bill might pass. First, they began to stress the need to study areas before they were designated as wilderness. Such studies would delay any potential withdrawals of land from the mining laws. Second, some mining groups testified that a wilderness bill would be acceptable as long as mineral exploration and development were allowed to continue.[31]

The Senate Interior and Insular Affairs Committee was ready to move forward with the bill but was delayed by personal issues within the committee (a death, an illness, and a retirement), the passage of the Multiple Use and Sustained Yield Act in 1960 (which recognized wilderness as a legitimate use of the national forests), and efforts to drastically alter the bill within committee. Before the committee could report out a bill, the Eighty-sixth Congress closed.

The Senate took up the issue again in February 1961, holding hearings on yet another wilderness bill (S. 174), introduced by Senator Clinton Anderson (D, N.Mex.), new chair of the Committee on Interior and Insular Affairs. The sections of the bill relating to mining were virtually identical to those most recently debated: in general no mining activity, but prospecting and mining could be undertaken with presidential authorization. Predictably, the mining industry opposed the legislation. The AMC continued to stress its themes of the lockup of minerals and the threat to national security. Sensing that the Senate would soon pass a wilderness bill of some type, the AMC also stressed the importance of allowing wilderness to remain open to mining activities if a wilderness bill were to pass.[32]

Other mining groups, however, became more strident in their opposition. The Northwest Mining Association claimed that "the wilderness bill is one of the most selfish, grasping, special-interest bills that has ever been presented to the Congress." The testimony continued: "This bill would play into the hands of our foreign enemies by discouraging discovery and production of metals necessary to national defense and necessary to a strong economy. This is no time for esoteric dilettantism; we must progress economically, or perish."[33]

The Senate committee, having done much of the groundwork in previous years, finally reported S. 174 out of committee in July. In the report accompanying the bill to the floor, the committee stated that "in view of the vast unexploited land areas of the Nation that remain and the safeguards written into S. 174, the majority of the committee does not feel that the mining industry will actually be injured by the bill." There was heated debate on the floor, including on the mining issue, but the bill passed easily, seventy-eight to eight, on September 6, 1961. Supporters of the wilderness bill had achieved victory in the Senate with only a minor concession on the mining issue.[34]

With the passage of S. 174 in the Senate, the focus shifted to the House, which had been holding off action until the Senate had passed a bill. In the House, mining interests and other commodity users had more influential friends, especially chair of the Committee on Interior and Insular Affairs, Colorado Democrat Wayne Aspinall. The House began its considerations on a wilderness bill (numerous bills had been introduced in the House, but the hearings focused on S. 174) with a series of three field hearings in Idaho, Colorado, and California. Mining interests focused their attention on achieving an exemption for mining in the House bill. They still opposed wilderness, but if a bill were to pass, they favored an amendment allowing mining to continue. Older themes, however, continued to be emphasized. Mining interests continued to stress their "rights" to prospect and mine on the public domain. The New Mexico Mining Association objected to S. 174 because "the well-established rights of every citizen under the mining laws would be completely wiped out."[35]

In May 1962 the House convened hearings in Washington, and again the mining interests were out in full force. It was becoming clearer that a bill would pass, so the strategy of the mining interests shifted: the mining community focused on making the wilderness bill more acceptable. They argued that mining and wilderness were not incompatible uses. Actual mining operations required very little space and would be hardly noticeable in wilderness areas. Also, abandoned mines would quickly return to wilderness conditions. In advocating leaving the wilderness open to mining, they indicated that they were willing to work under special restrictions established by the government (as long as they were reasonable). They favored review of lands before they were declared wilderness by the USGS or the Bureau of Mines and a periodic mineral review of the wilderness lands by these agencies. The Mining and Metallurgical Society of America offered a specific amendment to S. 174: "Anything in this Act to the contrary notwithstanding, lands within the wilderness system shall continue to be open to prospecting and subject to location and entry in the same manner and to the same extent as under existing mineral laws of the United States and the rules and regulations applying there to."

Mining interests also sought the addition of two procedural components in the wilderness bill. First, Congress, not the executive branch, should have sole power to create new wilderness areas. This provision was favored by commodity groups because they thought that their friends in Congress (e.g., Aspinall) would be less likely than the executive branch to propose more wilderness. And second, mining interests did not want national forest primitive areas to be automatically included in the wilderness system. They simply wanted to limit the acreage included at first as much as possible. In addition to making their case for what a wilderness bill should include, both the AMC and the Mining and Metallurgical Society continued to underscore the specific objections that they had already made to a wilderness system in general.[36]

Aspinall had promised that the House would pass a wilderness bill after the Senate did, and he kept his word. In June 1962, the House Committee on Interior and Insular Affairs began marking up a bill (H.R. 776) introduced by Representative John Saylor (R, Pa.) that included mining provisions identical to the Senate bill. In full committee, though, the entire text of Saylor's bill was replaced with an amendment offered by Aspinall. The new bill dealt not only with wilderness but also with overall public-lands management authority. In terms of mining, the bill that came out of subcommittee included a ten-year exemption for mining within wilderness. That is, the mining laws would continue to function until December 31, 1972. In full committee, however, an amendment offered by Representative John Chenoweth (R, Colo.) to allow wilderness areas to remain open for twenty-five years (through December 31, 1987) was adopted. The bill called for periodic mineral reviews of wilderness areas conducted by the USGS and/or the Bureau of Mines. Finally, the bill contained a provision mandating the review of all wilderness areas every twenty-five years to determine if the designation were still suitable.

Supporters of the wilderness bill were strongly opposed to the amended H.R. 776, which they regarded as an antiwilderness bill. Saylor argued, "The substitute being reported to the House by the Interior Committee is a perversion of the wilderness preservation legislation that so many conservationists and conservation agencies throughout the Nation have been advocating so long and so earnestly." The bill was reported out of committee on August 30, and Aspinall, under the instructions of the committee, sought to have the bill considered under a suspension of the rules. He was denied this by House Speaker John McCormack (D, Mass.). Aspinall did not seek new instructions from the committee, and the bill died with the Eighty-seventh Congress. Even though the bill they favored did not pass, mining interests had succeeded in blocking passage of a wilderness bill in the House. Meanwhile, wilderness supporters saw their Senate victory die.[37]

The Senate considered another wilderness bill in early 1963. The hearings were primarily a formality because the Senate had passed the bill so easily in the prior Congress. The mining community was there to testify in opposition to the bill (S. 4, identical to S. 174 as passed in 1961, that is, no mining without specific presidential authorization), arguing that an amendment exempting mining should be adopted.[38]

Despite this continuing pressure from mining interests, the Senate Committee on Interior and Insular Affairs held firm to the restrictive language. The committee leadership reasoned that the mining interests could make their case in the House, where they were likely to receive a more sympathetic hearing. An undiluted Senate bill would also allow for more flexibility in the likely conference committee. On the Senate floor, an amendment offered by Senator Peter Dominick (R, Colo.) to allow mining to continue in wilderness areas for fifteen years was rejected decisively. On April 9, 1963, the Senate passed the wilderness bill seventy-three to twelve, again with only a minor concession on mining in wilderness areas.[39]

With Senate passage, the burden shifted back to the House. The mining interests took hope from a comment made by subcommittee chair Walter Baring (D, Nev.), who promised the AMC no wilderness bill unless "proponents are willing to move in the direction of the compromise offered by the House committee last year [H.R. 776]." As in the last session, the House began consideration of the wilderness bill with a series of hearings in the West in early 1964. At these hearings, and later ones in Washington, the mining interests continued to state their opposition to the bill as passed in the Senate (S. 4). They focused their attention on H.R. 9162, introduced by Representative John Dingell (D, Mich.). The bill, worked out in advance by Aspinall and executive-branch agencies, included language that allowed all mining activities to continue in wilderness areas until December 31, 1973, a period of ten years.[40]

Most mining interests opposed this ten-year exemption as being too brief to be meaningful, but they almost universally preferred this to the alternative of immediate closure in the Senate bill. They also saw this ten-year exemption as an opening that could be extended. The typical pattern of testimony was to argue that mining should be exempted from the provisions of the wilderness bill generally, *but* if there were to be a wilderness system, H.R. 9162 was preferable to S. 4 or H.R. 9070, *provided* that H.R. 9162 be amended to allow for a twenty-five-year exemption for mining activities (as was the case in H.R. 776). This House bill allowed mining interests to get their foot in the door. The question became, could they get the door to remain open forever (as they hoped), and if not, could they get it to remain open longer? The mining interests also argued for regular, periodic reviews of the wilderness areas by the government, with provisions to allow mining if these reviews revealed significant mineral deposits.[41]

It appeared most likely that the House Committee on Interior and Insular Affairs would select H.R. 9162 as the wilderness bill to mark up. This bill was most favorable to commodity users and had been the result of negotiations by Aspinall and the executive branch. However, Aspinall chose H.R. 9070, the Saylor bill, as the bill to be marked up. The main rationale for this selection, it appears, was in respect of the death of Howard Zahniser of the Wilderness Society, perhaps the foremost proponent of the wilderness bill. A subcommittee amendment allowing for a twenty-five-year continuation of mining laws within the wilderness areas was adopted. This was one of the two major differences with the already passed S. 4 (the other being the requirement of congressional action to designate additional wilderness areas). The mining provisions withstood floor debate, and the House passed H.R. 9070 by an overwhelming vote of 373 to 1 on July 30.[42]

In the conference committee appointed to work out a compromise acceptable to both chambers, the mining exemption was one of the major issues to be dealt with. Preservationists were strongly opposed to any exemption, as were the Senate conferees. Senator Anderson, the lead negotiator, discussed the issue with the Forest Service. The agency pointed out that it had regulated mining in wilderness areas for the past twenty-five years and would continue to do so during any period that the wilderness areas were open to mining. The Forest Service also indicated that it would employ even stricter regulations to control mining in the wilderness areas if necessary. Representative Aspinall, chief negotiator for the House, also tried to assure Anderson that the mining interests did not envision permanent roads or mechanized access to carry out their work. Based on these assurances, Anderson and the Senate conferees acquiesced to the mining clause. The House compromised a bit by reducing the period that the mining laws would apply from twenty-five to nineteen years (i.e., wilderness areas would be open to mining laws through December 31, 1983). The final report adopted S. 4 with this change regarding mining and additional nonmining changes. Both chambers approved the conference report on August 20, and on September 3, 1964, the Wilderness Act, with a nineteen-year exemption, was signed into law.[43]

The act as finally passed excluded mining activity from wilderness areas in general, with a few specific exceptions. Foremost of these exceptions was that the mining laws would be allowed to function for nineteen years. It must also be stressed that mining could occur after December 31, 1983, on valid claims made prior to that date. The functioning of the mining laws, however, would be subject to regulations developed by the secretary of agriculture. These regulations would deal with things such as ingress and egress, facilities, waterlines, and so forth. The mining areas were also to be restored to the full extent possible. An additional restriction on the mining laws is that claimants could patent title only to the minerals, not to the land. Prospecting was to be allowed in wilderness areas, as long as it was "carried on in a manner compatible with the preservation of the wilderness environment." And finally, the USGS and Bureau of Mines were charged to conduct systematic, periodic mineral surveys of the wilderness areas and make these findings available.[44]

Although many mining interests were opposed to this final bill, arguing that the wilderness lands should remain open without a time limit, they had made a vast improvement over the bills first introduced, and the bill twice passed by the Senate. The supporters of mining interests could also claim a substantial victory in the nineteen-year exemption for mining. It soon became clear, however, that the victory of the mining interests was primarily a paper one.[45]

In 1966, Kennecott Copper indicated that it was planning to develop an open-pit copper mine in the Glacier Peak Wilderness area in Washington. The Agriculture Department and the Interior Department announced their opposition to the mine, the Forest Service indicated that it would place tight restrictions on the mine and related activities, local opposition groups formed, and eventually the Washington congressional delegation announced its opposition. Kennecott soon capitulated, announcing in 1967 that it would not develop the deposit.[46]

In the early 1970s, a court case regarding mining in the Boundary Waters Canoe Area (BWCA) in northern Minnesota further demonstrated that mining in wilderness areas would not be accomplished easily. George St. Clair, searching for low-grade copper-nickel in the BWCA, notified the Forest Service that he was going to begin drilling with heavy equipment. The Izaak Walton League filed suit to prevent the mining activity. In federal district court, Judge Philip Neville issued a permanent injunction against any mining in the BWCA. In his decision he stated: "To create wilderness and in the same breath to allow for its destruction could not have been the real Congressional intent, and a court should not construe or presume an act of Congress to be meaningless if an alternative analysis is possible." Although this decision was overturned at the appellate level, it sent a clear message to the mining industry.[47]

Since these decisions, the mining industry has not attempted any major development within wilderness areas. Rather, it has attempted to block additions to the system and to have areas of high mineral potential excluded from such areas. For example, the Central Idaho Wilderness Act of

1980 contained language stating that if economically minable deposits of cobalt were discovered in the wilderness, it could be mined under regulations that would apply to nonwilderness national forest lands. The inclusion of such language is a clear sign of the failure of the 1964 exemption to protect the mining option.[48]

Despite the de facto closure of the wilderness to mining, preservationists continued to attack the exemption as a cancer within the wilderness. The regulations established by the Forest Service have helped somewhat, but they have not prevented the continued scarring of wilderness and wilderness study areas. Also, the threat that existing claims will be developed remains. Mining interests also continued to stake claims in the wilderness, primarily to establish legal footholds.[49]

Although preservationists were generally successful in meeting their goal, they had to retreat in the face of an institutionalized idea in the mining policy regime. It must be stressed that mining was the only activity allowed potentially to expand and continue in wilderness areas, due primarily to the privileged position of economic liberalism institutionalized in the 1872 Mining Law. This privileged idea and the extralegal property-rights system that was institutionalized with it proved to be the key difference between mining and other uses.

This privileged idea did not escape unscathed, though. New claims would only be allowed for nineteen years, not indefinitely. And, despite what the law said, wilderness areas have been in reality closed to new mining due to strict regulatory requirements. In a sense, since the passage of the Wilderness Act, preservationists have been able to exert influence in other parts of the state. They have been able to make use of the separation of powers and a fragmented bureaucracy to achieve goals that eluded them in Congress. Hence, economic liberalism was institutionalized enough to garner special treatment for mining, but the Wilderness Act represented the first cracks in its foundations of privilege.

Strategic Minerals and the Public Lands, 1975–1985: National Security or Economic Revival?

The Arab oil embargo of 1973 led many analysts in the United States to focus on the potential for similar economic disruptions of hard-rock minerals. Analysts argued that the United States was import dependent, and hence vulnerable, for a large variety of these minerals. These analysts, and later the political community, focused most of their attention on the so-called strategic minerals, namely chromium, cobalt, and the platinum-group metals, which were of critical importance for military and industrial applications. Although the 1973 oil embargo helped to substantiate the claims being made that the United States faced a minerals crisis, numerous additional reasons existed as to why some feared severe mineral shortages.

First, there was a general fear that the world might be running out of certain mineral resources, a fear related to the computer projections made by the Club of Rome during this period. The lack of investment being made by mining corporations to develop new mineral sources was another cause for concern. Third, and most related to the oil embargo, were concerns about interruptions in trade due to import dependence. These interruptions could be due to the formation of cartels (similar to the Organization of Petroleum Exporting Countries [OPEC]), the use of embargoes (similar to the Arab oil embargo), war and civil disruption (which affected cobalt production from Zaire), and a resource war launched by the Soviet Union (which would entail Soviet control over the crucial resources of southern Africa). A final reason for concern was the problem of surges in demand due to military needs or the fluctuations of the business cycle. Underscoring these concerns, during this period prices for many of these minerals increased dramatically: from June 1978 to June 1979, the price of cobalt increased 253 percent; platinum, 59 percent; and nickel, 44 percent.[50]

The domestic mining industry was a strong voice among those warning of a potential minerals crisis. These industries had been in an economic slump, and they viewed the issue as a way to tie the improvement of their industry to national security (as they had attempted to do in the earlier debate about wilderness). The aspect of this issue that the domestic mining industry most focused on was increasing minerals production from the public lands. During the 1970s, a large percentage of the public lands had been withdrawn from the functioning of the 1872 Mining Law (and the 1920 Mineral Leasing Law). These withdrawals were due to the establishment of wilderness areas (which were technically open to the functioning of the law but de facto unavailable), temporary withdrawals for the study of additional lands for potential wilderness designation (Wilderness Study Areas) under the Forest Service's Roadless Area Review and Evaluation (RARE) I and II and the Bureau of Land Management's Federal Land Policy and Management Act (FLPMA) review, and the unsettled nature of Alaskan lands, on which legislation regarding state lands, Native American lands, and "national interest" lands (i.e., parks, wildlife refuges, national forests) was pending. Also, existing wilderness areas were to be closed to mineral leasing and new mining claims at the end of 1983.

Additionally, numerous other general environmental laws were passed

Table 1. Federal Public-Land Availability for Hard-Rock Mineral
Development, 1975

	Formally Closed (%)	Highly Restricted (%)	Moderate/Slight Restriction (%)
Non-ANCSA* lands	15.9	6.1	44.0
ANCSA* lands	18.0	—	16.0
Total	33.9	6.1	60.0
Total (millions of acres)	271.4	48.4	480.1

Source: Office of Technology Assessment, *Management of Fuel and Nonfuel Minerals in Federal Land*, pp. 337–38.

*ANCSA: Alaska Native Claims Settlement Act. These lands were temporarily withdrawn pending final legislation on Alaskan lands (passed in 1980).

throughout the 1960s and the 1970s that affected mining. These laws included the National Environmental Policy Act (NEPA), the Clean Air Act amendments, and the Clean Water Act amendments. These laws typically made hard-rock mining on the public lands more expensive and further hurt the economic condition of the domestic mining industry.[51]

Prior to these various withdrawals (ca. 1968), only 17 percent of public lands were withdrawn from the mining laws. A report by the Department of the Interior Task Force on the Availability of Federally-Owned Mineral Lands concluded that in 1974, mineral location under the 1872 Mining Law was prohibited on 42 percent of the federal lands, was severely restricted on 16 percent, and was moderately restricted on 11 percent of the lands. Another set of figures contained in an Office of Technology Assessment (OTA) report done in 1979 (see table 1) indicate less land closed to mining entry. This report also concludes that of the non–Alaska Native Claims Settlement Act lands formally closed to hard-rock mining, only about one-third of the public lands were closed for cultural or environmental reasons (e.g., national parks). The remaining two-thirds were closed for other uses (e.g., military bases, irrigation projects, petroleum reserves).[52]

In the discussions on strategic minerals from 1975 to 1985, there was much debate as to which figures were correct. As the Interior Department and OTA reports suggest, there was significant confusion even within government. For interest groups, the confusion was just as great, if not greater, because the figures were out of date and subject to constant revisions (following the Alaska Lands Act, for instance). Typically, the industry cited figures with the highest possible amount of land closed to mining, a figure commonly ranging from 67 to 73 percent. Environmentalists cited lower figures, often in the 40 percent range, and stressed the importance of nonenvironmental withdrawals.

In the mid-1970s, the mining industry, supported by the general business community, cited this combination of withdrawn lands, increased environmental regulations, and the strategic minerals threat to push for increased access to the public lands. If the domestic mining industry had access to these lands, they could produce all the minerals needed for national security and economic growth. For example, a 1973 volume assembled by the Society of Economic Geologists reported the theme that would run throughout this issue for the next twelve years: "In recent years there has been a move to deny to mineral exploration large areas of public lands through (1) creation of wilderness areas; (2) the use of provisions of the National Environmental Policy Act (NEPA) to delay public land lease sales and drilling permit applications; (3) cancelling of or refusal to grant mineral patents; and (4) (difficult to document) harassment and intimidation by government officials at many levels. . . . Exploration and development of mineral deposits on public lands will continue to meet political resistance and to face higher costs."[53]

Following the Arab oil embargo, these more general concerns could be placed within the rubric of national security and strategic needs. In 1974 and 1975, articles in *Forbes*, *Nation's Business*, and *Science* dealing with strategic minerals indicated that the issue was on the nation's political agenda. In 1975, J. Allen Overton, president of the American Mining Congress, claimed that the United States was "heading straight toward a self-imposed malnutrition of minerals."[54]

In this debate, the visions of the public interest of economic liberalism (and technocratic utilitarianism) favored increased access for private corporations; the preservationist conception of the public interest opposed any further access. The AMC, chief representative of the mining community, argued tirelessly for increased access to potential minerals on the public lands. By opening more of the public lands for mineral development, a significant portion of the strategic minerals problem could be solved; any minerals developed in the United States would decrease import dependence and vulnerability. This would also have economic benefits for the nation through more jobs and a lower foreign debt. This perspective clearly reflects economic liberalism: the best way to manage the resources on the public lands is to transfer those lands to the private sector, where the free market will take over. If we have a strategic minerals crisis, allow mineral exploration and development on the public lands and the free market will solve the problem. There is nothing the market can do, however, if the lands continue to be unavailable.[55]

The supporters of preservationism were largely fighting a defensive battle on this issue. They sought to protect lands they had succeeded in withdrawing from mineral entry. Mining in such areas would deprive society of scenic views, wildlife habitat, and generally unspoiled areas. Preservationists claimed that no strategic minerals crisis existed, and that even if it did, it could not be solved through the public lands (most of which were open for exploration, and which did not contain many of the minerals most in need). Preservationist groups argued that the entire issue of strategic minerals was a scheme to help big business.

Despite these opening salvos, the mining interests had a tough battle on their hands with preservationists over the continuation of mining in national parks. Six national parks or national monuments had remained open to hard-rock mining when they were established. Preservationists were especially concerned with mining in Death Valley and Glacier Bay National Monuments and started a campaign to convince Congress to pass a law closing these six areas to future mining. Despite the efforts by mining interests, such a law was enacted in September 1976. Clearly, the strategic minerals argument had not yet won over Congress.[56]

The issue, however, had become important to some members of Congress, especially Representative Jim Santini (D) of Nevada, chair of the Subcommittee on Mines and Mining of the Committee on Interior and Insular Affairs. In February 1977 Santini drafted a letter advocating the development of a comprehensive minerals policy by the executive branch and the appointment of a minerals adviser to the president. Forty-two other representatives, mainly from the West, signed the letter that was delivered to President Jimmy Carter. The president met with a group of these representatives at the White House in June and later in the year announced the establishment of a major interagency review of nonfuel minerals policy. Further evidence of congressional interest is demonstrated by hearings held in 1977, 1978, and 1979 by the Senate Committee on Commerce, Science, and Transportation and the House Committee on Science and Technology on materials and minerals policy.[57]

Over the next few years, the strategic minerals—public lands availability question was subsumed within the larger debate over Alaskan lands. Mining interests (along with other commodity and development interests) fought to minimize the number of acres that would be declared wilderness or designated national parks or wildlife refuges; preservationists fought to maximize the acreage. Typically, the case to leave these lands open to minerals was based on general development/economic growth. The need to develop more domestic sources of strategic minerals was used as a supporting argument. For example, AMC president Overton argued: "Disjointed and uncoordinated government policies have resulted in almost three-fourths of all public lands being withdrawn from mining and exploration for minerals. . . . Now there is a bill in Congress to place OFF LIMITS a staggering 102 million acres of public lands in Alaska—thus undoing Seward's foresight by Washington's folly. . . . While Alaska has not been fully explored—and never will be, if this bill passes—we know it is a treasure house of oil, gas and coal and very many other strategic minerals." A typical environmental response downplayed the importance of minerals on these 102 million acres of land; one environmental response stated that more than 75 percent of Alaska's minerals are outside these lands according to the state's computerized resource inventory.[58]

Despite the Alaska issue, Congress continued to focus some attention on strategic minerals. In 1979, the congressional pace regarding strategic minerals began to accelerate, culminating in the passage of the National Materials and Minerals Policy, Research and Development Act of 1980. The bill began in the House Science and Technology Committee, and was passed by the House in December 1979. The House Committee on Interior and Insular Affairs was also busy in 1979, holding a series of hearings on nonfuel minerals policy. These hearings were an oversight of the Carter administration's nonfuel minerals policy review, announced in late 1977. At these hearings, members of the Mines and Mining Subcommittee, chaired by Santini, expressed their displeasure over the tardiness of the policy review and the lack of vigor with which the administration was undertaking the review.[59]

The following year, the same House Mines and Mining Subcommittee printed a report on the policy implications of United States minerals vulnerability. Among the topics covered was public-lands access. The report criticizes the government for its failure to manage these lands to benefit minerals production: "Despite these repeated recommendations and the long-recognized fact that the public lands of the West and Alaska hold great promise for mineral discoveries, government continues to restrict or prohibit their use for this economic and strategic national good. The most deplorable aspect of this short-sightedness is that it is being done without any real knowledge of the losses involved, without any attempt to understand long-term impacts, and without any government accountability for the consequences." Also criticized is the alleged antimineral bias in the

Interior Department: "The orientation of the Department of the Interior, the Department solely responsible for the development of the mineral resources of the public lands, has been one of fundamental skepticism if not outright opposition to mineral exploration and development." Lastly, minerals on the public lands are viewed as a key component of economic growth and national security: "Public lands provide the United States the opportunity to significantly offset foreign mineral dependence, to decrease a growing balance of payments deficit, to create jobs and to play a role in the reindustrialization of the Nation. Yet that will only take place if these lands are available for mineral exploration, development and production. It is therefore critical that the Department of the Interior put to an end its opposition to mineral use of the public's land."[60]

The report offered four specific recommendations regarding the public lands: (1) the Interior Department should generally make lands more available for mineral development; (2) mineral exploration and development in wilderness areas should be encouraged under the provisions of the Wilderness Act; (3) the mining law exemption in the Wilderness Act (allowing new claims through the end of 1983) should be extended through 2000, and any new wilderness areas should be open to Mining Law claims for twenty years after designation; and (4) "Mineral values of public lands should be placed on a priority at least equal to the environmental concept of 'areas of critical environmental concern' and other similar classifications." Many of these recommendations were part of the Minerals Security Act introduced by Representative Santini in 1981.[61]

The Senate was also in action in 1980, with two committees holding hearings on minerals policy. The Energy and Natural Resources Committee held hearings specifically on H.R. 2743, the National Materials and Minerals Policy, Research and Development Act, the bill passed by the House in December 1979, and the Commerce, Science, and Transportation Committee held field hearings in Nevada and New Mexico. In Nevada, a state senator argued that the major problem facing mining was "the threat of additional wilderness in areas that are yet to have adequate exploration." The executive director of the Nevada Mining Association sounded a similar alert: "The presence of these withdrawn areas, these wilderness areas, is going to have a major impact, a negative impact on the development of Nevada's resources." Similar complaints were made in New Mexico, with the wilderness withdrawals made under RARE I and RARE II, the BLM wilderness review, and the Alaska lands situation drawing strong criticism.[62]

In October 1980, the Senate considered and passed the Materials and

Minerals Policy bill that had passed the House in December 1979, and the bill became law. The act did not have much substance; it was general in nature, advocating a more coherent and coordinated minerals policy in the United States. The law, which makes numerous references to the need to address critical minerals necessary for national security, declares it to be the "policy of the United States to promote an adequate and stable supply of materials necessary to maintain national security, economic well-being and industrial production with appropriate attention to a long-term balance between resource production, energy use, a healthy environment, natural resources conservation, and social needs." Clearly, this is not a very specific (or controversial) policy declaration. It reflects the act well. The president, in recognition of the fragmentation within the state, was directed to "coordinate the responsible departments and agencies" and submit to Congress a program plan within one year describing how the goals within the act would be achieved. This report was delivered to Congress by President Ronald Reagan in April 1982.[63]

In November 1980, the supporters of a stronger national defense, of less environmental regulation, and of greater mining interest (and other commodity) access to public lands won a victory far larger than the Materials and Minerals Policy Act. President Reagan, his appointees, and a Republican-controlled Senate meant that mining interests generally, and especially those focusing on the strategic minerals problem, would have strong supporters of economic liberalism and their specific interests in positions of power. New Senate Energy and Natural Resources Committee chair James McClure (R, Idaho) said that "the committee's oversight would be aimed specifically at speeding up the decision-making process among those who manage Federal lands." McClure, sympathetic to the strategic minerals issue and a supporter of demands for increased access to the public lands, also suggested that greater access to public lands for fuel and nonfuel minerals would be forthcoming under the Republicans.[64]

The new secretary of the interior, James Watt, also did not disappoint mining interests with his early economic liberalism rhetoric. Upon taking office, Watt promised to allow increased access to the public lands for mining interests: "Strategic mineral dependence on foreign sources could best be averted by opening federal lands for mineral exploration and development." He continued, "We must allow the private sector the opportunity to explore the mineral potential on public lands.... We cannot have a healthy policy unless we have access to public lands." He concluded: "As minerals manager of the public's lands, I will oppose single use designation of those lands if there is evidence that their withdrawal means a significant loss of fuel or nonfuel mineral resources vital to our economy and the nation's interest. . . . [The Interior Department will speak for] the very real public interest involved in the protection and preservation of a strong minerals sector."

In addition to his general position on mining on the public lands, Watt also made his skepticism about the lack of mineral exploration in wilderness clear: "We have put into wilderness about 80 million acres of land, and we've never properly inventoried them. . . . We don't know what minerals are there, we don't know what energy potential is there, and right now the United States is vulnerable to a natural resource attack or war. We're vulnerable to a strategic mineral shortfall to such a degree that we're now dependent upon Russia and Southern African nations to meet our needs."[65] Watt was quick to support his words with actions. In July, the Interior Department directed the National Park Service to revise its regulations to allow mining on five national recreation areas (NRAS) it managed. The revisions were made, and by December 1981 mining was allowed on the 400,000-plus acres in the NRAS. Part of Secretary Watt's justification in taking this action was to reduce United States dependence on imported strategic minerals.[66]

Due to this change in administrations, the strategic minerals issue rose on the congressional agenda as well. In testimony before the Senate Energy and Natural Resources Committee in early 1981, the AMC testified that "we regard the denial of access to the Federal lands as the paramount public land mineral issue facing us today. . . . The trend of withdrawal of large acreages of land is inimical to our national interests." The group went on to propose the indefinite postponement of the December 31, 1983, closing of wilderness areas to the Mining Law.[67]

Mining advocates in Congress, led by Representative Santini, introduced H.R. 3364 (with forty cosponsors), the National Minerals Security bill of 1981, in April. With a new, more sympathetic administration in place, it was time for mining interests to try to achieve the goals that eluded them in the 1970s. The bill was designed to deal with the major general concerns of the mining industry, as identified in the National Materials and Minerals Policy Act, and to deal with the issue of strategic minerals more specifically. The major components of the bill dealt with the creation of a Council on Minerals and Materials, methods to increase minerals production from the public lands, improved mineral data collection and analysis, revisions in the tax laws to favor mining, regulatory reform to favor mining, a strengthening of the national defense stockpile, a review of antitrust legislation unfavorable to the mining industry, and a requirement for the secretary of state to submit "an annual report on the foreign policy of the United States as it relates to the availability of minerals for domestic use."[68]

Title III of the bill, Domestic Mineral Resource Potential, dealt specifically with mining on the public lands. The policy section makes the aims of the title clear:

(a) The discovery of new sources of mineral wealth from the public domain is in the best interest of the national economy and security. It is in the public interest to foster, encourage, and promote the exploration and development of domestic mineral resources.

(b) The purpose of this title is to provide the means by which the Secretary of the Interior . . . may make available for mineral location and leasing under applicable Federal law those public lands heretofore withdrawn, classified, restricted, or closed to such purposes.

The bill directed the secretary of the interior to review all land-use plans developed under FLPMA to consider their effect on mining, to determine the acreage withdrawn or closed in any way to the mining laws, and to request from private persons or firms nominations for closed lands to be reviewed for potential mineral exploration and development. After a review process, the secretary is "authorized and directed" to apply the general mining laws to these lands. The final, and most controversial, aspect of the title concerned wilderness areas. Section 306 of the bill amended the Wilderness Act to allow wilderness areas to remain open to claims under the Mining Law of 1872 (and the Mineral Leasing Law) until December 31, 1993, a ten-year extension.

Two additional components of the bill applied to public-lands policy. First, the bill would create the Council on Minerals and Materials in the Executive Office of the President, which would be responsible for the coordination of minerals policy and would balance the existing Council on Environmental Quality. And second, the USGS would be charged to examine and classify the public lands regarding their mineral potential, especially "strategic and critical minerals."

Representative Santini's Subcommittee on Mines and Mining held hearings on the bill in October 1981. The positions taken by the various witnesses were predictable. The AMC, Chamber of Commerce, Minerals Exploration Coalition (a group of exploration geologists), and the Rocky Mountain Oil and Gas Association all favored the public-lands sections of the bill, with one exception: they supported leaving wilderness areas open to the mining laws indefinitely, rather than for the ten additional years specified in H.R. 3364. The environmental groups that testified at the hearings—Audubon, Friends of the Earth, National Wildlife Federation, Public Lands Institute/Natural Resources Defense Council, Sierra Club, and Wilderness Society—all opposed the bill. They argued that the public-lands provisions of the bill were unnecessary, that mining interests were overstating the amount of public lands closed to mining, and that the strategic minerals issue was a false one. More specifically, they were opposed to leaving wilderness areas open to the mining laws for an additional ten years, the increased discretionary power for the secretary of the interior to open lands closed to mining, and the mineral council, which they argued would be the equivalent of mining lobbyists at taxpayer expense and would place mineral use above other resource uses of the public lands.[69]

During this period of legislative policy debate, Secretary Watt continued to use his executive powers to open more public lands for minerals access. He focused his attention on wilderness areas, claiming that the Wilderness Act not only allowed but mandated exploration and development of both hard-rock and fuel minerals on these lands. Watt also favored a twenty-year extension of the mining exemption in wilderness areas, allowing them to remain open through 2003. Furthermore, if Congress did not act within a specified period of time on wilderness study lands managed by the Forest Service and the BLM (two and six years), these lands should be released for multiple-use activities.[70]

Using his executive authority, Watt planned to begin issuing oil and gas leases on wilderness areas, something no other interior secretary had done. In May 1981, Congress acted to prevent him from issuing such leases in the Bob Marshall Wilderness in Montana. Shortly after, in September 1981, the BLM issued leases for oil and gas exploration in the Capitan Mountain Wilderness in New Mexico. Congress reacted by threatening to close all wilderness lands immediately (rather than waiting until the end of 1983) via legislation. To ward off such action, Watt announced a moratorium on leasing activity through December 30, 1982 (after the 1982 congressional elections). Because Watt had no discretionary control over hard-rock mining claims, this moratorium only applied to oil and gas leasing.[71]

In February 1982, on national television, Watt seemed to completely change his position on mineral development in wilderness areas. He said that he now supported withdrawing all wilderness and wilderness study areas from mineral leasing laws and location under the 1872 Mining Law until 2000. A bill to implement the policy, the Wilderness Protection Act (H.R. 5603), was introduced by Representative Manuel Lujan (R, N.Mex.). The bill would withdraw from the mining laws all wilderness and wilderness study areas through the year 2000. During that time, private exploration would be permitted if it did not harm the wilderness, the president would have the power to open the lands for mineral development if he found an "urgent national need," and government agencies would study the lands for mineral potential. Wilderness study areas could only be studied for a specific period, after which Congress could either designate the lands as wilderness or they would revert to multiple-use lands. And lastly, the entire national wilderness system would be reviewed by Congress during the 106th Congress (when the mineral withdrawal ends). Environmental groups strongly opposed this bill due to the presidential loophole and the need to reauthorize all wilderness in 2000. A Wilderness Society representative called it "a wilderness destruction bill."[72]

The House considered both H.R. 5603 and H.R. 5282, introduced by Representative Philip Burton (D, Calif.), which would have immediately and permanently closed wilderness and wilderness study areas to mineral leasing and mineral location. Lujan and Burton constructed a compromise bill, H.R. 6542, which passed the House in August 1982. This bill closed wilderness to oil and gas leasing, moving up the December 31, 1983, deadline, but did not affect hard-rock minerals. The Senate considered, but did not pass, a similar bill (S. 2801). Although a law was not enacted, Congress used appropriations legislation in fiscal years 1983 and 1984 to prevent the Interior Department from issuing any oil and gas leases in wilderness areas. Congress also passed a Florida wilderness bill, vetoed by President Reagan, that would have prevented the issuance of phosphate leases unless the president declared them needed to prevent a domestic shortage.[73]

In the midst of this debate over minerals and wilderness, President Reagan, in April 1982, submitted his minerals report and program to Congress (as directed by the Materials and Minerals Policy, Research and Development Act of 1980), which reflected the themes of economic liberalism. One of the four major themes in the report was the need for increased access to the public lands for mining (the others were research and development, data collection, and stockpiling). "This national minerals policy recognizes," the report begins, "the vast, unknown and untapped mineral wealth of America and the need to keep the public's land open to appropriate mineral exploration and development." This knowledge is to be learned through the private sector: "New mineral deposits will not be found unless the private sector looks for them. It is to the nation's advantage to encourage this search." Several of the specific public-lands recommendations contained in the report were similar to provisions in H.R. 3364.[74]

The report endorsed Watt's Wilderness Protection Act, called for an acceleration of the secretary of the interior's review of public lands closed to mining to determine if these withdrawals continue to be necessary, and endorsed the public identification of "Areas of Critical Mineral Potential" on public lands for priority withdrawal review. At Senate hearings to review the report, Interior Secretary Watt again stressed the importance of increased access to the public lands for mining purposes: "For various reasons, the Federal Government has refused proper access to our public lands owned by all the American people that could stimulate the industry and allow us to meet many of our strategic minerals and materials needs. . . . We think that if we are going to prepare for national defense, improve the quality of life in America, and enhance our environment as well as create jobs, we have got to change that, and we are in an aggressive manner doing that."[75]

Testifying at these hearings, groups representing mining interests greeted the report in a lukewarm fashion. They favored the overall theme of the report: increasing access to the public lands. For example, the AMC commended the administration "on the positive steps that it has taken on reviewing and revocation of many obsolete public land orders"; and the National Association of Manufacturers "supports efforts to increase the availability of federal land for mineral exploration and development." These groups, however, strongly opposed the withdrawal of wilderness lands through the year 2000. "We do not believe that the proposed action is in the national interest," the AMC testified. The American Institute of Professional Geologists argued that "at this time of serious shortages of strategic minerals, Secretary Watt's proposal to lock-up these lands and their minerals represents a distinct threat to the national economy and military security of all Americans." The Minerals Exploration Coalition also strongly opposed this policy, arguing instead for leaving wilderness areas open to mining indefinitely.[76]

Environmental groups responded negatively to all components of the report dealing with public lands. They responded that the plan was yet another aspect of the administration's general plan to turn over more public resources to private enterprise. A Sierra Club representative argued that "the land availability issue is a red herring. . . . We could open every national park in the country and not alleviate our minerals shortage. We know what minerals are on the public lands." The plan was characterized by the Wilderness Society as "a further ripoff of the public lands and a sweetheart deal for industry."[77]

A more thorough review of the report by environmental groups argued that the president's report was based on at least three faulty assumptions. First, the report states that 40 to 68 percent of federal public lands is closed to mineral exploration and development. Environmentalists counter that the figure is 38 percent, including much land closed for nonenvironmental reasons. Second, the report "grossly exaggerates the threat posed to the nation's security by dependence on foreign suppliers of strategic minerals." And third, the implication that the United States is not self-sufficient in strategic minerals due to federal policies harmful to domestic mining is wrong: "Research shows that many strategic minerals are not produced in the United States because they are simply not found here in measurable quantities or it is uneconomical to mine them."[78]

During the flurry of activity regarding Secretary Watt, mineral leasing in wilderness areas, the various wilderness protection bills, and the president's report on minerals policy, Congress continued to consider the National Minerals Security bill, and an additional bill, H.R. 4281, the Critical Materials Act. In 1982, the House Armed Services Committee held two hearings on the latter bill; the House Committee on Science and Technology also held hearings on the bills (as well as oversight of the Materials and Minerals Policy Act of 1980) during April 1982. At these hearings, the environmental community offered its most complete statement on strategic minerals and the public lands. In a report prepared by the Environmental Policy Center, Friends of the Earth, National Audubon Society, National Wildlife Federation, Natural Resources Defense Council, Sierra Club, and Wilderness Society, these groups presented their rationale opposing plans to increase mineral access to the public lands. According to this report, "less than 28% of the federal lands are withdrawn from claim staking and less than 22% from leasing for environmental reasons." The report concludes: "There is no compelling reason to open public lands which are not already accessible to mineral development. Our public lands belong to all Americans." In later testimony by the Audubon Society, the group argued that based on "the factors in the strategic minerals debate that acceleration of mineral development on public lands or weakening the regulations which provide protection for surface resources on these lands are unnecessary actions, and are, in fact, not in the public interest."[79]

Although the administration's policy regarding wilderness had become mired down in legislative and public debate, its policy to make more withdrawn lands available to mineral development continued through executive actions. On December 27, 1982, Interior Secretary Watt announced that he had removed 800,000 acres of BLM lands from wilderness study area classification. The following types of lands were removed: (1) all nonisland parcels under 5,000 acres in size, (2) all split-estate lands, and (3) lands being considered for wilderness due to the wilderness characteristics of adjoining land (rather than wilderness characteristics on the study lands). All of these lands, except the split-estate lands, were opened to access under the 1872 Mining Law. Watt also declared that he planned to review an additional 5 million acres for potential removal from withdrawals.[80]

In July 1984, a version of the National Security Minerals bill was enacted. The new law, the National Critical Materials Act, was a much watered down version, leading only to the creation of a National Critical Materials Council. The act began with the usual general platitudes before noting executive-branch fragmentation on minerals policy: "There exists no single entity with the authority and responsibility for establishing critical materials policy and for coordinating and implementing that policy." In an effort to deal with this and other problems, the law establishes the National Critical Materials Council within the Executive Office of the President, and authorized its existence through 1990. The three members are to be appointed by the president and are to serve the role as coordinators of federal critical-materials policy. Note that the controversial sections of H.R. 3364 regarding withdrawal powers and a lengthened mining exemption in wilderness areas have been removed from the bill. It is, like the 1980 Materials and Minerals Policy Act, just a set of general policy statements and the creation of an advisory council. In February 1984, the secretary of the interior created an additional advisory board, the National Strategic Materials and Minerals Program Advisory Committee. This board, designed to gain private support for administration programs, consisted of persons from the private sector, primarily business interests.[81]

The executive branch continued to do what it could to open more of the public lands to mining. From 1980 through 1985, the BLM reclassified two hundred million acres of public lands. Approximately thirty million acres of these lands were opened to the operation of the mining laws. The AMC was "generally pleased by the reclassifications," whereas the National Wildlife Federation found the changes to be "very damaging to the public interest." At about the same time, the BLM withdrew a new set of regulations dealing with the exercise of mining rights in wilderness areas. The new rule "was perceived by many to violate our policy to protect wilderness areas." After the public storm in 1982 regarding mining and wilderness, the administration wanted to prevent any appearance of the lack of a commitment to protect wilderness.[82]

Despite the administration's actions to open more public lands to mining, some in Congress continued to be displeased with the executive branch's overall commitment to strategic minerals. For the second time in four years, a congressional oversight hearing expressed disappointment that the president was not implementing an act with sufficient vigor (the National Critical Materials Act). Although the administration was working to increase access to the public lands, its congressional critics charged that it was not developing a coherent, coordinated critical-materials policy.[83] The issue of strategic minerals began to dissipate on the national agenda by the mid-1980s as many of the potential fears failed to materialize. Mineral cartels as successful as OPEC did not form—in fact, OPEC's stature as a cartel diminished greatly over the period; a Soviet resource war failed to materialize; and, the United States was more, not less, import dependent on strategic minerals in 1985 than it had been in 1975.[84]

The focus of the strategic minerals issue was to gain increased access to the public lands by reducing the acreage covered by withdrawals and by keeping wilderness areas open for minerals access. Mining interests were successful, to some degree, in achieving the first goal, but not the second. Primarily via executive action under the Reagan administration, millions of withdrawn acres were made available for entry under the 1872 Mining Law. Efforts were made to keep the wilderness open but were met with strong opposition, and wilderness areas were closed to further claims and oil and gas leasing on December 31, 1983, as scheduled. Hence, what success mining interests had came through the executive branch rather than through Congress.

In retrospect, the public-lands component of the strategic minerals issue is best viewed as an issue of economics rather than of national security. The depressed mining industry saw the issue as a vehicle to gain increased access to the public lands. These previously unavailable areas might hold new discoveries that could help revive the industry. This interpretation is further supported by noting that the key congressional leaders in the effort to open more of the public lands were from mining states (e.g., Santini and McClure). Environmental groups made similar claims (that the issue was an economic one) and were moderately successful in framing the issue. Supporters of neither the economic liberal nor the preservationist perspective were able to dominate the issue. Rather, both gained some victories and suffered some defeats. Economic liberals were successful in reducing the amount of land withdrawn from mineral development but unsuccessful in their efforts to open wilderness areas to mineral development. Preservationist victories and defeats were the converse.

Throughout this issue, the fragmentation of the state in the hard-rock mining policy regime is apparent. Nearly all the major actors (e.g., Congress, environmentalists, mining interests, OTA) were unified in calling for a more coordinated executive branch in the mining policy regime. Among the chief aims of the two laws passed during the period were the reduction of this fragmentation and the development of an executive-branch capacity to develop and implement coherent policy in regard to strategic minerals (and minerals in general). Under the Reagan administration, the major actors on strategic minerals policy were the Department of the Interior (including BLM, Bureau of Mines, Minerals Management Services [MMS], Office of Minerals Policy and Research Analysis, and USGS), Department of Commerce, Department of Defense, Federal Emergency Management Agency (FEMA), General Services Administration (GSA), Annual Materials Plan (AMP) Steering Committee, Committee on Materials (COMAT), and the Cabinet Council on Natural Resources and Environment (CCNRE). The working group for Reagan's 1982 minerals report included many more actors, including nine cabinet departments: Agriculture, Commerce, Defense, Energy, Interior, Justice, State, Transportation, Treasury, and the CIA, Council of Economic Advisors, EPA, FEMA, GSA, NSC, OMB, Office of Policy Development, Office of Science and Technology Policy, and U.S. Trade Representative. If the sheer number of executive branch agencies involved in this policy is not enough evidence of fragmentation, a vivid example is supplied by the government's lack of knowledge as to how many of its acres had been withdrawn from entry under the 1872 Mining Law. The differing figures contained in Interior Department and OTA studies underline the confusion. A further example of poor management was the government's loss of millions of dollars in mineral revenues (primarily from oil and gas leasing) due to poor management. Yet another agency, the MMS, was created to collect these revenues, but it has not received high marks for its efforts. Despite the passage of the Materials and Minerals Policy Act of 1980 and the National Critical Materials Act in 1984, the subsequent creation of the National Critical Materials Council, or the calls for the creation of a cabinet department of minerals and mining and for the appointment of a minerals adviser to the president, this fragmented state in the hard-rock mining policy regime has remained.[85]

Fundamentally, though, the debate was shaped by the institutionalization of economic liberalism in the 1872 Mining Law. According to a number of analysts, it is because this law is still the basis of hard-rock mineral policy on the public lands that land withdrawals are prevalent: it is the only way to control what happens on the land. Once a claim is staked or a patent is granted, federal control over the land virtually ceases. Although the Forest Service and BLM have issued regulations concerning surface resources and mining claims, this is a recent development. Also, neither the Forest Service nor the BLM has proved itself a vigilant enforcer of these regulations thus far. The miners still do not need a permit to proceed with their work, which could have required that certain conditions be followed. Without this control, the federal government often simply withdrew the land from entry rather than risk unacceptable harm from mining. Altering the 1872 Mining Law could reduce the need for withdrawals and allow greater access to the public lands for mining interests.[86]

In many ways, the property-rights system legitimized by the 1872 Mining Law was the foundation on which the issue of strategic minerals–public-lands access was debated. The law recognized that prospectors could obtain private property rights on public lands by staking claims and following mining codes. The federal government had virtually no say in this process as long as the land was open to the functioning of the 1872 Mining Law. Hence, the only way to prevent or control mining in a given area of land was to withdraw it from the 1872 Mining Law. The system of property rights that controlled hard-rock mining on the public lands was one of extremes: mining with virtually no control, or no mining. There was no room for the government to establish conditions for mining. This lack of flexibility led to increased withdrawals and restrictions, which led to mining industry complaints about a lack of access. With a more flexible system of property rights, it is likely that less land would have been withdrawn, and perhaps this issue would not have arisen, a point stressed by those who favor replacing the 1872 Mining Law with a leasing system.

The strategic minerals and public-lands issue demonstrates the continued institutionalization of the 1872 Mining Law. This law, reflecting the privilege of economic liberalism, continued to define the hard-rock mining policy regime. And a weak, fragmented state in this regime reflected the continued pattern of a privatized policy. Because this law was still in place, the only option to prevent mining was to withdrawal the land from the workings of the act. To mining interests, this violated their rights to the public land and was counter to the public interest. To supporters of preservationism, such withdrawals were necessary. The issue dissipated in an ambiguous manner: the law remained, the mining interests received access to some withdrawn lands (but not as much as they wanted), and wilderness areas remained closed. The privilege of economic liberalism was still in place, but it was badly tarnished.

Conclusion

Hard-rock mining policy on the public lands has been based on the 1872 Mining Law for more than 120 years. Although there have been minor changes in the law and the policy, it remains the centerpiece of this privatized policy pattern. This law institutionalizes the economic liberal perspective of the public interest regarding public-lands management, the unchallenged mode of thinking about the public lands at the time. As time passed, however, and different conceptions of the public interest developed, the 1872 Mining Law remained firmly in place. A technocratic utilitarian challenge to the law never developed because, in the hard-rock mining realm, the technocratic utilitarians were allied with the economic liberals. They believed, and continue to believe, that the market should guide mineral development. A preservationist challenge to economic liberalism did develop. Examinations of both the passage of the Wilderness Act and the strategic minerals issue demonstrate the clashing of the different conceptions of the public interest. Economic liberals argued that wilderness should remain open to private sector development and later argued for a reduction in the amount of land closed for mineral development. When this land was made unavailable, it prevented the free market from operating to achieve the public interest. Preservationists argued that some of the public lands must be preserved from exploitative uses, including mining, in both instances.

In the case of the Wilderness Act, the institutionalization of the 1872 Mining Law and the privilege of economic liberalism that it incorporates were strong enough to prevent preservationists from excluding mining from wilderness areas. Indeed, mineral claims could be made for another nineteen years. No other commodity interest was able to achieve this same legislative victory. Despite the failure to ban mining legislatively, preservationists have been successful in keeping mining out of wilderness via the executive branch. Executive branch withdrawals, coupled with the lack of mining in wilderness, were central to economic liberal complaints about lack of access to the public lands for strategic minerals. In this case, the preservationists were able to protect wilderness areas, but many other areas were reopened to mineral access. The weak state capacity in this policy regime, the result of more than 125 years of a privatized policy pattern, was of central importance in this issue.

Overall, then, the 1872 Mining Law and economic liberalism has been institutionalized in the hard-rock mining policy realm. Once this idea became embedded in the policy regime, it served to shape hard-rock mining policy, reflected in the privatized policy pattern. Although preservationists and their ideas have been able to make changes at the edges of the policy, they have been unable to make any central ones. Although the interests of both these groups, the mining industry and preservationists, are clearly at stake, it is this pair of ideas of what serves the public interest that is the keystone of their arguments, the historical thread that connects past policy battles with present and future ones. As George Julian, a chief congressional opponent of the 1866 Mining Law wrote in 1883, the "misfortune of this legislation is heightened by the probability of its continuance; for it is not easy to uproot a body of laws once accepted by a people, however mischievous their character. . . . [This] wretched travesty [of a law may be] permanently engrafted upon half a continent." Thus far, Julian has been right, at least in respect to the ability of the act's supporters to resist change.[87]

MANAGING THE
NATION'S FORESTS

THE MARKET, SCIENCE,
OR MORALITY?

In this chapter, I examine federal forestry policy on the public lands from the late 1800s through 1985. The development of the professionalized policy pattern and the institutionalization of the idea of technocratic utilitarianism will be demonstrated in three case studies. The first case is the formation of the forestry policy regime, from the mid-1880s to 1910. During this period, the forest-reserve system was established and the Forest Service gained the authority to manage these lands. The idea of technocratic utilitarianism became embedded within the agency and became privileged in the policy regime, providing the foundations for the professionalized policy pattern. The remaining two cases examine significant challenges to this policy pattern. The passage of the Wilderness Act was an important incursion to the privilege of technocratic utilitarianism, as supporters of preservationism were able to alter the professionalized pattern on the issue of wilderness. The privatization initiative of the 1980s demonstrated that technocratic utilitarianism was still privileged enough to ward off the challenge of supporters of economic liberalism. In the 1990s, technocratic utilitarianism is still the privileged idea that grounds the professionalized policy pattern, but both the privilege and the pattern have become considerably weaker.

Formation of the Forest Service:
Incorporating Scientific Management
into the Federal Government

In the final half of the nineteenth century, interest in the nation's forests increased due to concern with their effects on watersheds and weather, perceived timber shortages, and the establishment of professional forestry in the United States. In 1873 the American Association for the Advancement of Science (AAAS) appointed a committee to make Congress and state legislatures aware of the importance of protecting forests and to advocate the introduction and passage of legislation to achieve this end. In 1875, the American Forestry Association (AFA) was formed to promote forestry and forest protection. Congress responded to these growing concerns, and in 1876 the federal government commissioned the first report on the condition of the forests in the country, to be undertaken by the Department of Agriculture. Subsequent reports were produced in 1878, 1880, and 1882, all of which recommended the creation of a forest-reserve system. In 1881 the Division of Forestry was established within the Department of Agriculture, which Congress statutorily recognized in 1886.[1]

Beginning in 1876, members of Congress regularly introduced bills to create forest reserves on the public lands. It was not until 1891, however, that such a bill passed. During this fifteen-year period, the main advocate of the forest reserves was the AAAS and the scientific community it represented. For instance, in 1890 the AAAS sent President Benjamin Harrison a message, calling upon him to "investigate the necessity of preserving certain parts of the present public forest area as requisite for the maintenance of favorable water conditions" and stating that pending such an investigation, all forest lands owned by the United States government should be withdrawn from sale. Although these scientists were successful in gaining presidential support (Harrison was won over by the AAAS message of 1890), they could not convince enough members of Congress to support their proposal. The opponents favored the status quo—economic liberalism: the public interest would be best served by transferring all the public lands to the private sector. This approach was supported both because it served the economic interests of these groups and because the idea was part of their larger ideology. The idea of government ownership of forest reserves, conversely, ran directly counter to this idea, and less directly, counter to their interests.[2]

In spite of this opposition, and due to some legislative maneuvering in a conference committee, the president received the power to establish forest reserves in the 1891 General Revision Act. When the bill went to conference committee, it contained no mention of forest reserves. The AFA, though, viewed this conference committee as an opportunity to push the forest-reserve measure through Congress. It determined that four of the six conferees favored forest reserves and that the remaining two members were not opposed. The AFA persuaded Interior Secretary Noble to ask the conferees to include a rider in the bill authorizing the president to establish forest reserves. The conferees were amenable and included such a rider in the bill as Section 24.[3]

President Benjamin Harrison created fifteen reserves with a total of more than 13 million acres, and President Grover Cleveland expanded the system to nearly 40 million acres by creating fifteen additional reserves. This expansion, coupled with the lack of management on the forest reserves, stimulated great protest, primarily in the West—western senators introduced three different bills to revoke the forest-reserve system. These opponents thought the land was being locked up and that this did not serve the public interest. Much of the problem was due to the language of the 1891 law, which did not allow for management of these lands.[4] Scientists, foresters, and commodity interests all wanted to see management allowed on the forest reserves. In 1897, an amendment to the Sundry Civil Appropriations Act allowed for the management and use of resources from the forest reserves. This lessened some of the opposition to the reserves from those who wished to graze livestock on them or harvest timber from them. The opposition returned in full force, though, during the administration of Theodore Roosevelt. Between 1901 and 1909 he reserved more than 148 million acres in more than one hundred new forest reserves. This was more than enough for western members of Congress; in 1907 they revoked the president's authority to create new reserves in six western states. Roosevelt further infuriated his opponents by establishing twenty-one new reserves just before signing the bill. This marked the end of the tremendous expansion of the forest reserves in the West, which constitutes the vast majority of land in the current national forest system.[5]

From the mid-1890s on, a variety of interests that were the chief supporters of making additions to the forest reserves coalesced around two ideas, technocratic utilitarianism and preservationism. Supporters of technocratic utilitarianism were those interested in retaining forest lands in government ownership so that they could be managed scientifically, by foresters, for the greatest good of society. The supporters of this idea—foresters and other scientists, groups such as the AAAS, the AFA, and the Society of American Foresters (SAF)—were able to convince others that scientific forestry, practiced by the government, would further the public interest. Also, the technocratic utilitarians were able to point to the waste, mismanagement, and danger inherent in private management. Historical evidence supported them, and they took that evidence to Congress and the public. Expert, not democratic, control was the aim of the technocrats.[6]

Technocratic utilitarianism's emphasis on expert control was squarely within the "search for order" constellation of ideas that historians have identified as the motivating theme of reform during the period. Society was becoming more interdependent due to urbanization, industrialization, and the rise of large corporations. These changes created substantial dislocation and confusion. In the forests the turmoil was apparent in ravaged stands of trees, fires, and floods. Technocratic utilitarianism promised to bring order to the forests. This idea, with its reliance on professional foresters, is part and parcel of a bureaucratic impulse that was widely viewed as the chief alternative to the status quo at the time. As Robert Wiebe writes, the bureaucratic alternative was based on "trained, professional servants [who] would staff a government broadly and continuously involved in society's operations. In order to meet problems as they arose, these officials should hold flexible mandates." He argues further that "each new administrative power . . . was predicated upon the continuous, expert management of indeterminate processes." The successful incorporation of these ideas is perhaps most clearly seen in the forestry policy regime.[7]

The supporters of technocratic utilitarianism were not alone in their battle against the seeming waste and disorder that accompanied the dominant economic liberalism idea. They were joined by those who argued that preservation of certain parts of the public lands was in the public interest. Indeed, the preservationists were far more numerous than those advocating technocratic utilitarianism. At this juncture they shared with the technocratic utilitarians the dismay over the condition of the forests left to private hands and the desire to have these lands retained by the federal government.[8]

As the forest reserves were expanded, the need for some management authority on the lands increased. As noted above, the 1897 Sundry Civil Appropriations Act included an amendment (now referred to as the Forest Service Organic Act) that allowed for management of the forest reserves. Despite the law, a further difficulty remained. The Division of Forestry was in the Department of Agriculture, the forest reserves were administered by the General Land Office (GLO) in the Department of the Interior. (At this time, all public domain lands were under the authority of the GLO, the main function of which had been and continued to be administering the disposal of the public lands.) Technocratic utilitarians began a campaign to have the administration of the forests transferred to the Agriculture Department, where foresters could manage the lands. Once these reserves were under the control of the Division of Forestry, they would be managed efficiently and scientifically by a well-organized, scientific agency, for "the greatest good to the greatest number for the longest time."[9]

The leading advocate of this transfer was Gifford Pinchot, head of the Division of Forestry beginning in 1898. In order to gain the transfer, Pinchot had to overcome the opposition of the GLO and commodity interests who were comfortable with the GLO's nonmanagement approach. With the use of a highly skilled publicity machine, the political support of President Roosevelt, and the support of the AFA and preservationists, he was able to engineer the transfer in 1905.

The publicity efforts of the agency deserve special attention. The natural constituency of the Bureau of Forestry (a name change in 1901) was small: the few foresters in the country, those advocating scientific management, irrigation interests. The bulk of those concerned with forestry issues were opposed to the forest reserves and the Bureau of Forestry. To deal with this, the agency sought to educate the public themselves. As Pinchot said, "Nothing permanent can be accomplished in this country unless it is backed by sound public sentiment. The greater part of our work, therefore, has consisted in arousing a general interest in practical forestry throughout the country and in gradually changing public sentiment toward a more conservative treatment of forest lands." The Bureau of Forestry (and later, the Forest Service) did not simply report information to the public, but rather, "campaigned aggressively to create public support for [his/its] vision of utilitarian forestry."[10]

The Bureau of Forestry used publicity and lobbying to help lessen opposition and gain support for the plan to transfer the forest reserves to the agency. Pinchot wrote numerous articles that appeared in such popular magazines of the day as *Century*, *Outlook*, and *Scientific American*. The bureau sent out handouts on its activities and on why the transfer made sense to numerous newspapers, magazines, and individuals. Rather than letting Congress or the public decide on the fate of the transfer issue, the bureau orchestrated a large-scale lobbying scheme designed to convince Congress and the public of the virtues of the transfer. The bureau did not abandon these tactics following its success in the transfer battle; rather, it continued to publicize its activities in order to bolster its position with Congress. A press office was created in 1905, and in fiscal year 1908, the Forest Service published 4.4 million copies of 220 publications. The agency's mailing list reached 750,000 by 1909, a list that included editors, reporters, community leaders, and professional leaders. By using this publicity machine, the Forest Service worked to shape public opinion in a way conducive to its goals.[11]

In 1901, Representative John Lacey (R, Iowa), chair of the House Committee on Public Lands, introduced two different bills authorizing the transfer. One bill did not make it out of committee, the other was bogged down with amendments and eventually failed in the full House. The majority of the members of the committee favored the bills, but there was strong opposition from certain western representatives and from Representative Cannon (R, Ill.), chair of the Appropriations Committee. Public Lands Committee members opposing the transfer filed a minority report in which they raised the objections of inefficiency and interdepartmental friction but did not discuss perhaps their main objection: the fear of more restrictive grazing rules on the forest reserves if they were managed by the Bureau of Forestry.[12]

Following the defeat of the transfer bill in 1902, Representative Franklin Mondell (R, Wyo.), an opponent of the Lacey bill, introduced a transfer bill in the House. Drawing on this newfound western support (due to Bureau of Forestry efforts to convince skeptical members of Congress, railroad interests, lumber interests, mining interests, and grazing interests that the forest reserves would be more fully open for commercial use under its management than they had been under GLO management), the bill passed the House. Even Joe Cannon, now Speaker of the House, chose not to waste his energies on holding up a transfer act with western support, even though he still opposed it.[13]

The transfer bill continued to encounter difficulties in the Senate. Senator Warren (R) of Wyoming introduced a bill authorizing the transfer, but the bill was blocked due to the opposition of South Dakota senator Alfred Kittridge (R), who blocked action on the bill in 1904 because the Homestead Mining Company, one of the most important companies in his state, feared that the transfer of the forest reserves to the Bureau of Forestry would threaten its supply of lumber for mining purposes from the Black Hills forest reserves. The following year, the bill was modified to address these concerns; it cleared both chambers of Congress, and on February 1, 1905, President Roosevelt signed the law and the Bureau of Forestry received the authority to manage the forest reserves.[14]

This act did not just transfer the forest reserves to the jurisdiction of the Bureau of Forestry; it also stipulated that all receipts from the forest reserves be placed in a special fund for forest management, controlled by the secretary of agriculture, for a period of five years. This fund was to be employed for the purposes of protecting, administering, improving, and extending the reserves. Fueled primarily by grazing fees, this fund grew to approach the agency's congressional appropriation by fiscal year 1906 ($800,000 versus $1,195,218). The agency estimated that if the special fund were continued, it would need no appropriations after 1911. The fund thus allowed the Bureau of Forestry a significant degree of autonomy as it provided funds with minimal congressional interference.[15]

On the day the Transfer Act was signed, Pinchot received a letter from Agriculture Secretary Wilson outlining the principles under which the forest reserves were to be managed. This letter was based upon language in the Organic Act authorizing the secretary of the interior to make any rules and regulations necessary for the proper management of the reserves. Nothing in the letter was a surprise to Pinchot because he wrote it and simply passed it on for Wilson's signature. This letter, along with the 1897 Organic Act, provided the general framework for policy on the forest reserves until the passage of the National Forest Management Act in 1976. Among the most important passages from the letter: "In the administration of the forest reserves it must be clearly borne in mind that all land is to be devoted to its most productive use for the permanent good of the whole people, and not for the temporary benefit of individuals or companies. All the resources of the reserves are for *use*, and this use must be brought about in a thoroughly prompt and businesslike manner, under such restrictions only as will insure the permanence of these resources . . . and where conflicting interests must be reconciled the question will always be decided from the standpoint of the greatest good for the greatest number in the long run." This letter was a further source of bureaucratic autonomy for the Bureau of Forestry. Numerous management decisions concerning the forest reserves were made without congressional authority, including many decisions Congress would not have sanctioned.[16]

The Forest Service (yet another name change in 1905) proved worthy of the management task. From 1899 to 1905, its budget had grown from

Table 2. Division of Forestry/Bureau of Forestry/Forest Service, Budget and Employees, Fiscal Years 1899–1908

Fiscal Year	Budget ($)	Number of Employees
1899	28,520	54
1900	48,520	123
1901	88,520	192
1902	185,440	247
1903	291,860	297
1904	350,000	307
1905	439,873	939
1906	1,195,218	1,390
1907	1,825,319	2,012
1908	3,572,922	2,753

Source: U.S. Department of Agriculture, *Annual Reports of the Department of Agriculture: 1900* (p. 103), *1902* (p. 386), *1904* (p. 317), *1906* (p. 642), *1907* (p. 62), *1908* (p. 775).

$28,500 to $439,873, its number of employees from 54 to 939. Both budget and employees grew further once management of the forests began (see table 2). As discussed above, the idea of technocratic utilitarianism pervaded the small and interconnected profession of forestry, through the SAF, educational institutions, and the Forest Service, and this unified profession also helped strengthen the capabilities of the agency. Additionally, the entire Forest Service staff was placed under civil service requirements, insulating the agency from some of the political pressures of the day. This was unusual during the period, when most jobs were still under the spoils system. An emphasis on efficient and scientific management (apparent in the *Use Book*, which specified policies to be followed in specific instances), and an esprit de corps (apparent in the standard badge and uniform, which the agency has had since 1906) among its employees were successful. In 1906, the Committee on Department Methods declared the Forest Service so much more efficient than other government agencies that it based its recommendations for other federal agencies on Forest Service practices.[17]

The combination of the special forest-reserve fund, the management discretion based on the 1897 Organic Act, the small, interconnected profession of forestry, the agency's public affairs program, and the placement of the vast majority of employees under civil service guidelines all contributed to the Forest Service's autonomy, an autonomy guided by technocratic utilitarianism. In establishing forest-reserve management authority in the Forest Service, the weak capacity of the GLO (due to corruption and the lack of professionalism) was avoided and the state fragmentation in this policy regime was overcome. The Forest Service was an island of capacity and autonomy in a generally weak state.[18]

As with the creation of the forest reserves, Congress soon acted to close the window of autonomy of the Forest Service. According to President Roosevelt, the Forest Service was under more attacks than any other government agency: "The opposition of the servants of the special interests in Congress to the Forest Service had become strongly developed, and more time appeared to be spent in the yearly attacks upon it during the passage of the appropriations bill than on all other Government Bureaus put together. Every year the Forest Service had to fight for its life." Gifford Pinchot similarly stressed the numerous attacks by "special interests" upon the Forest Service: "It is the honorable distinction of the Forest Service that it has been more constantly, more violently and more bitterly attacked by the representatives of the special interests in recent years than any other Government Bureau. These attacks have increased in violence and bitterness just in proportion as the Service has offered effective opposition to predatory wealth. The more successful the Forest Service has been in preventing land-grabbing and the absorption of water power by special interests, the more devious, and the more dangerous these attacks have become."

The separate forest-reserve fund was eliminated in 1907, Forest Service publicity efforts were ordered stopped by President William Taft in 1909, Pinchot was forced to resign in 1910 due to a controversy with Interior Secretary Richard Ballinger, and interest-group politics became the political arena for the Forest Service. Throughout the remainder of the century, the Forest Service has sought to maintain its bureaucratic autonomy in managing the national forests, but it has been constrained by interest-group pressures in a manner in which it was not constrained in the early 1900s.[19]

Those who coalesced under the economic liberalism idea of the public interest opposed the forest reserves. They did not oppose all reservations or all government ownership. Certain lands reserved for national parks, such as Yellowstone or Yosemite, might be acceptable. But the massive reservation of lands containing economic resources was counter to the public interest, and not coincidentally, their own. Those who opposed these withdrawals were those who wanted to use (or had been using) resources from these reserves: settlers, speculators, miners, timber harvesters, cattlemen, sheep ranchers, and others who might be involved in these or related ventures. Government forest reserves would constrain their ability to prosper and society's as well. There were also those who supported the forest-reserve system as long as it remained under the control of the GLO because these arrangements were conducive to the economic benefit of particular firms and sectors of the economy. They opposed the transfer because they had established a working relationship with the GLO and were skeptical about the Bureau of Forestry.

Although the preservationists favored the transfer of the forest reserves, the differences between their idea and technocratic utilitarianism became more apparent as Forest Service management continued. The preservationists did not suggest that all forest reserves should become parklike preserves from which resources could not be used. Rather, they argued that certain lands of exceptional beauty and quality should be preserved. Such an approach clearly runs counter to the idea of technocratic utilitarianism, which is concerned with resource use, not resource preservation. Although the preservationists and technocratic utilitarians began to battle over a number of issues after 1900, it was the battle over Hetch Hetchy reservoir in Yosemite National Park, from 1908 through 1913, that was the decisive one. The technocratic utilitarians successfully supported the flooding of the valley to supply water to San Francisco, leading to a bitter split with the preservationists. But before this split, during the crucial turn-of-the-century period, these groups were allies.

The creation of the forestry policy regime at the turn of the century was the result of the interplay of interests and ideas. Interests supporting government forest reserves coalesced around the ideas of technocratic utilitarianism and preservationism, were victorious over those interests most intent on obtaining material gain from these lands, interests that coalesced around the idea of economic liberalism. The forestry policy regime that resulted from this political interplay featured two critical characteristics: a dominant idea and a strong and autonomous state agency. Even though the autonomy of the Forest Service declined following 1907, the agency enjoyed a period of relative autonomy through the 1960s. The capacity of the agency remained strong, and technocratic utilitarianism continued to be the privileged idea guiding the agency.[20]

The Wilderness Act of 1964: Preservationist Blow to Technocratic Utilitarianism

The idea of wilderness has been important and influential throughout American history—in terms of literature, philosophy, and sociology. In the early history of the United States, wilderness was a place to be conquered, a place to be contrasted with civilization. An additional important way to think about wilderness surfaced in the 1860s and 1870s. In 1864, the Yosemite Valley and Mariposa Big Tree Grove were transferred from the federal government to the state government of California to be preserved for their scenic beauty. Eight years later, the United States established the first national park in the world—Yellowstone, reserved "as a public park or pleasuring-ground for the benefit and enjoyment of the people." These were the first government actions designed to preserve large areas of natural lands for aesthetic, moral, and recreational values. The national park system continued to develop, often with the strong support of commercial interests who would benefit from such parks, especially the railroads. It is important, though, to keep in mind that these parks did not wholly preserve land areas. Rather, hotels and roads were often established in the parks in order to make them more amenable to visitors.[21]

A more pristine wilderness became the focus of discussions in the Forest Service during the 1920s. Employees Aldo Leopold and Arthur Carhart advocated establishing large areas in the forests to be preserved in their natural character, free from roads and resource exploitation. Eventually, the Forest Service responded by designating a portion of the Gila National Forest in New Mexico as wilderness in 1924, by supporting wilderness management plans developed by Carhart for the Superior National Forest canoe areas in northern Minnesota in 1926, and by establishing regulations to deal with wilderness on a national scale in 1929. These L-Regulations, as they were called, described procedures for establishing "primitive" areas in national forests. The regulations were vague and ambiguous because, among other reasons, many foresters strongly opposed any designations that would prevent the use and management of resources. Also, because "primitive" areas could be established by administrative action, they could similarly be abolished by administrative action. Hence, the protection of these areas was quite tenuous.[22]

Over the next decade, the Forest Service faced numerous land-management challenges from the National Park Service. As the NPS sought to expand the national park system, it often looked to national forest lands as the raw material for new parks. These efforts to expand via the national forests were frequently successful: Olympic National Park (1938), Rocky Mountain National Park (1939), Glacier Bay National Monument (1939), King's Canyon National Park (1940), and Jackson Hole National Monument (1943) were all either created from or expanded onto national forest lands. This loss of land concerned the Forest Service: as an organizational entity it was losing resources and power. Some in the Forest Service felt that this NPS threat would continue unless the Forest Service could demonstrate a viable recreation and scenic area management program. Robert Marshall, a strong supporter of wilderness and an employee of the Forest Service (he was appointed head of the Recreation and Lands Division in 1937), argued that a more rigorous wilderness program on the national forests would be just such a program.[23]

In 1935, the Wilderness Society was formed by Leopold, Marshall, and a small group of others concerned with wilderness. The group was organized "for the purpose of fighting off invasion of the wilderness and of stimulating . . . an appreciation of its multiform emotional, intellectual, and scientific values." This group, the first dedicated solely to furthering the idea of wilderness and a wilderness system, went on to play a key role in the battle for the passage of the Wilderness Act. The Wilderness Society, after passage of the act, continued to be the chief preservationist group concerned with wilderness.[24]

Using the NPS threat as leverage, Marshall and other wilderness proponents were able to convince the Forest Service to adopt a new set of regulations. The U-Regulations, adopted in 1939, established three new classifications of wild land. "Wilderness" areas were to be at least 100,000 acres in size, were to be designated by the secretary of agriculture, and were to be without roads, motorized transportation, timber harvesting, or occupation. "Wild" areas would be from 5,000 to 100,000 acres in size and were to be under the same management restrictions as "wilderness" areas with the exception that they would be designated by the chief of the Forest Service rather than by the secretary of agriculture. Lastly, the regulations created "recreation" areas that were to be of 100,000 or more acres and were to be managed "substantially in their natural condition." In such areas, road building and timber harvesting, among other activities, could take place at the chief's discretion. Lands that had been classified as "primitive" under the L-Regulations were to be reviewed and reclassified under the U-Regulations.[25]

This reclassification did not take place quickly or smoothly. Many of the difficulties can be traced to the death of Robert Marshall in 1939. With the chief proponent of wilderness in the Forest Service dead, those large factions both inside and outside the Forest Service who opposed wilderness and preservation as the unnecessary lockup of resources could more readily slow any efforts to give wilderness more protection. This lethargy in the Forest Service did not go unnoticed. In the early 1950s, preservationist groups decided that it was time to gain statutory protection for wilderness. They had never been entirely happy with the tentative protection offered wilderness by administrative designation, and the tortoise-paced reclassification further suggested that the Forest Service could not be trusted to protect the wilderness. Additionally, the Forest Service could not prevent mineral development or power development on these lands through administrative designation. And finally, a controversy over wilderness land management in Oregon further convinced preservationists that administrative protection was not enough. In reclassifying the Three Sisters Primitive Area in 1957, the Forest Service classified only 200,000 acres of the 253,000-acre area as wilderness. The remaining 53,000 acres, much to the dismay of preservationists, were opened to timber harvesting.[26]

The idea and goal of a national wilderness system was perhaps the primary focus of the entire preservationist movement. The core of this idea was the need to preserve large tracts of nature untouched and unspoiled by humans. The movement had originally focused its practical political efforts on national parks; preserving the jewels of the North American continent. By the 1920s, however, their focus had begun to shift. The national park system and the NPS were established; many of the nation's most beautiful areas had been preserved. But the NPS development within these parks (e.g., hotels, restaurants, roads) detracted from preservationist objectives. Preservationists began to focus more attention on the Forest Service and its protection program. The campaign for a wilderness statute was based on the argument that neither the Forest Service nor the NPS were adequately preserving these areas.

Howard Zahniser, director of the Wilderness Society, led the campaign for a wilderness law, beginning in the late 1940s. He introduced the idea at the Sierra Club's First Biennial Wilderness Conference in 1949. At the second such conference in 1951, Zahniser formally proposed working for the adoption of such a law. Any further efforts to gain a wilderness law were put on hold, however, while the entire preservation community united to fight the proposed Echo Park dam.[27]

Briefly, the Echo Park dam was part of a Colorado River Storage Project proposed in the 1940s. If built, the dam would have flooded significant portions of canyons within the Dinosaur National Monument in Colorado. The preservation movement viewed this proposal as a critical threat to what they had achieved in terms of protecting nature. If a dam could flood this national monument, no national park, no national monument, no wildlife refuge, no wilderness area would be safe. Many viewed the battle as a modern equivalent to the fight over Hetch Hetchy in the early part of the century. As discussed in Chapter 2, the flooding of Hetch Hetchy Valley in Yosemite National Park was a tremendous blow to the preservationist movement and served to split the technocratic utilitarians and the preservationists, who had up until that time been allies in efforts to maintain certain lands in government ownership. The Echo Park battle lasted until 1956; the preservationists were successful. They had used all of their resources to defeat the administration, the Interior Department, and western development interests. When the Colorado River Storage Project was statutorily approved in 1956, without the Echo Park dam, "the American wilderness movement had its finest hour to that date."[28]

After achieving victory in the Echo Park dam issue, preservationists returned their attention to a wilderness bill. In addition to the concerns discussed above, preservationists hoped to draw upon the positive public opinion they had generated in the Echo Park campaign. Zahniser repeated his plea for a national wilderness preservation system in 1955 at the National Citizen's Planning Conference on Parks and Open Spaces for the American People. Less than a week after Zahniser's speech, Senator Hubert Humphrey (D, Minn.) had the speech reproduced in the *Congressional Record*. Having found an interested friend in Congress, preservationists went to work to produce a draft bill. Cooperating in the drafting of the bill were such organizations as the Citizens Committee on Natural Resources, the Council of Conservationists, the Federation of Western Outdoor Clubs, the Izaak Walton League, the National Parks Association, the National Wildlife Federation, the Sierra Club, Trustees for Conservation, the Wilderness Society, and the Wildlife Management Institute.[29]

On June 7, 1956, Senator Humphrey introduced S. 4013, which called for the establishment of a national wilderness preservation system. Four days later, Representative John Saylor (R, Pa.) introduced an identical bill in the House (H.R. 11703). The original Senate bill, which had nine cosponsors, sought to establish a wilderness system consisting of lands in national forests, national parks, national monuments, wildlife refuges, and Indian reservations. Within these wilderness areas certain activities would not be allowed: farming, logging, grazing, mining, building roads, using motorized vehicles. No new management agency was to be created; the areas would be managed by the agency currently managing the lands. All "wilderness," "wild," and "recreation" areas in the national forests were to be immediately incorporated into the system. "Primitive" areas were to be reviewed to determine which specific areas merited inclusion in the system. Based on these classifications, approximately fourteen million acres of national forest lands might be designated as wilderness. The bill also specified areas in forty-nine national parks and twenty wildlife refuges for inclusion in the system. Wilderness areas on Indian reservations were to exist only with the cooperation of tribal councils. Expansion (or reduction) of the system would come primarily by executive prerogative: the executive branch would make the proposed change known to Congress, and if neither chamber passed a resolution opposing the change within 120 days, the alteration would take effect. The last major component of the bill was the creation of the National Wilderness Preservation Council, an advisory board to consist of government officials and citizens appointed by the president.[30]

Eight years would pass before, in 1964, a wilderness bill would be signed into law by President Lyndon Johnson. During this time, sixty-five different wilderness bills would be introduced in either the House or the Senate, thousands of pages of testimony would be collected at thirteen Senate hearings and five House hearings held in Washington and throughout the West, and preservationists would have to make significant concessions to the original Humphrey bill.[31]

Because both the Humphrey and Saylor bills were introduced relatively late in the second session of the Eighty-fourth Congress, wilderness proponents did not have much hope that a bill would pass, or even that hearings would be held, before the end of the session. They did, however, want to get the issue onto the agenda and into the realm of public visibility. With the wilderness bill on the agenda, coalitions supporting and opposing the legislation formed quickly, and at the hearings held in 1957, the composition of these coalitions became clear.

Senator Humphrey and Representative Saylor reintroduced wilderness bills in February 1957. Senator James Murray (D, Mont.), chair of the Committee on Interior and Insular Affairs, became a cosponsor of Humphrey's bill (S. 1176), paving the way for quick hearings. Hearings on the bill were held in Washington on June 19 and 20, and a broad array of witnesses were on hand to present testimony. The composition of the supporting and opposing coalitions represented at these hearings, with a few exceptions, remained intact throughout the battle for passage of wilderness legislation.[32]

The coalition supporting wilderness legislation consisted of a variety of groups, some mentioned above, dedicated to activities such as preserving nature and natural systems, protecting the environment, protecting wildlife, and hunting and fishing. The prowilderness coalition employed a variety of reasons and arguments to support its position. These reasons all fit within the overall idea that the preservation of wilderness was good for the kinds of ethical, instrumental, moral, and spiritual reasons discussed in Chapter 2. Throughout the political debate, coalition members focused on more specific rationales in addition to this broader philosophical position. First, they argued that wilderness should be preserved for future generations: wilderness was an important part of the American heritage and our descendants should be able to enjoy this heritage. It is also unfair to foreclose the options of future generations; because we do not know what the interests or desires of future generations will be, we should err in the conservative direction—conserve the options for future generations. Wilderness advocates also stressed that it was an important time to act to save wilderness due to growing scarcity: wilderness itself was becoming scarcer, as were certain natural resources. As these natural resources grew even scarcer, economics would dictate the exploitation of sources in wilderness areas that were once deemed too costly. Hence, wilderness needed protection now, before the areas were spoiled due to economic forces.

The prowilderness coalition cited the problems of administrative protection discussed above. Namely, that wilderness areas were still vulnerable to mining operations and water development, and that the Forest Service was doing a less than adequate job of protection in some instances. To critics who argued that wilderness violated the multiple-use concept of forest management, prowilderness spokespersons countered that wilderness was quite compatible with multiple-use management. In a wilderness area, the forest would be used for recreation, watershed protection, and wildlife habitat. Only timber harvesting and mining would not be allowed. And lastly, the coalition emphasized the relatively small amounts of land to be declared wilderness. The vast majority of lands—total and public— would be open for commodity use.

At the first Senate hearings on wilderness legislation in 1957, the groups testifying in support of the bill were the American Planning and Civic Association, the Citizens Committee on Natural Resources, the Defenders of Furbearers, the Federation of Western Outdoor Clubs, the Izaak Walton League of America, the National Parks Association, the Nature Conservancy, the Sierra Club, the Trustees for Conservation, and the Wilderness Society. Additional groups that were part of, or became part of, the prowilderness coalition included the AFL-CIO, the American Camping Association, the Audubon Society, the Garden Clubs of America, the National Speleological Society, and the National Wildlife Federation. Throughout the course of the eight-year debate, numerous local and regional organizations also testified in favor of the wilderness bill or made public their support of the bill. These groups ranged from the New York State Conservation Council to the Carolina Mountain Club to the Minnesota Conservation Association to the editor of *Field and Stream* magazine in Idaho.

The prowilderness coalition, in contrast to the antiwilderness coalition, was united in its goals and desires. Its purpose is captured in an editorial appearing in the *Living Wilderness* in 1957: "American conservationists today are the vanguard of what surely must become a program in perpetuity. The tenseness of our responsibility and opportunity is in our necessity to fashion wisely a policy and program that will successfully keep the wilderness forever wild. We could miss this opportunity. We could fail. We could be forced to retreat. We could become the rearguard of an inevitably disappearing resource. But we are not that now. . . . We are working for the future." In the 1957 hearings, David Brower of the Sierra Club summarized the position of the coalition on future generations: "Thomas Jefferson, long ago, said that one generation could not bind another; each had the right to set its own course. . . . But deeds are not matching words. This generation is speedily using up, beyond recall, a very important right that belongs to future generations—the right to have wilderness in their civilization, even as we have it in ours."[33]

The leading opponents of the bill were in many ways predictable. Supporters of both economic liberalism and technocratic utilitarianism were opposed to such a statutory wilderness system. Representing economic liberalism were the groups and organizations who feared that the wilderness system would lock up valuable resources, resources that these concerns needed to function. It is hard to predict the future and our resource needs, they argued; hence it is unwise to lock up any of the resources on our public lands. This is especially true with mineral resources. The mineral potential of many proposed wilderness areas was unknown, and the idea of locking up these places before we knew what resources were in them struck many as ill advised, especially those in the mining industry. The resources, mineral and otherwise, might be necessary to continue or stimulate economic growth, both in the West and throughout the country, and certain minerals might be needed for strategic reasons. There was a need to maintain flexibility and multiple use.

This opposition coalition used these issues of economic development and national security to further their claims that they were not just looking out for their own narrow economic interests, but rather were looking out for the best interests of the country as a whole. Relatedly, they argued that wilderness proponents represented a very small segment of American society, and that these wilderness proponents were acting selfishly in their efforts to lock up these resources. Most people did not seek or care to seek a wilderness experience: they wanted resources developed; they wanted to be able to drive to wild places in their cars and sleep in their trailers when going to such places for recreation. A final argument made by those opposing wilderness legislation was that such legislation was unnecessary. The present system of national parks and national forests protected areas and provided an adequate and flexible system of wild lands.[34]

The chief sources of opposition were interests concerned with grazing, mining, motorized recreation, timber harvesting, and water development. More specifically, in this first set of hearings opposing groups included the American Mining Congress, the American National Cattlemen's Association, the American Pulpwood Association, the Interstate Stream Commission, the National Association of Engine Boat Manufacturers, the National Lumber Manufacturers Association, and the Upper Colorado River Association. Throughout the process, additional groups also made their membership in the opposition coalition clear, the most important being the American Petroleum Institute, the Farm Bureau, the National Reclamation Association, and the United States Chamber of Commerce. In addition to these more nationally oriented groups, numerous regional and local groups involved in commodity use or supportive of unlimited economic development opposed the legislation (e.g., the Inland Empire Multiple Use Committee, the Western Colorado Cattleman's Committee, and the Western Resources Conservation Council). Finally, specific corporations and companies that were members of trade associations (i.e., specific forest-products companies or specific mining companies) often made their opposition to wilderness legislation known (e.g., Boise Cascade Corporation and Kennecott Copper Corporation).

To get a flavor for the positions and arguments made by these opposition groups, it is useful to examine some of the testimony offered at the numerous wilderness hearings. At the 1957 Senate hearings, A. Z. Nelson of the National Lumber Manufacturers Association offered testimony on behalf of the timber industry: "When commercial resources are locked up our economy is deprived of additional tax dollars, pay envelopes, and consumer products. Many, therefore, are deprived of economic sustenance so as to provide a very limited number of individuals with wilderness pleasures." At the 1962 House hearings, W. Howard Gray of the American Mining Congress testified:

Our military strength and economic growth are dependent upon the full utilization and development of the productivity of our public lands through private enterprise.

We believe that the public interest is best served by keeping the public domain open for the discovery and mining of minerals, both metallic and non-metallic. No area of public domain should be closed to prospecting and mining. . . . We specifically oppose the so-called wilderness bill, and condemn recent actions of the U.S. Forest Service in requesting withdrawal of large acreages of land for the specified purpose of precluding the location of mining claims.

Gray continued his testimony by comparing "the objectives of the wilderness bill to the U.S.S.R."[35]

Technocratic utilitarians had been consistently skeptical of a wilderness system, a skepticism that varied from hostility to acquiescence. From Gifford Pinchot through the present, they have been opposed to the idea of locking up resources. They want to manage resources scientifically, so that the resources can be exploited in the most efficient and productive manner possible. Wilderness preservation represents a waste of resources. This antipathy could be seen in the opposition of many foresters to the L-Regulations of the 1920s and the U-Regulations of the 1930s. This lockup of resources was one of the two main themes of technocratic utilitarian opposition to wilderness throughout the entire debate.

The second main theme was that such a wilderness system would infringe upon the management discretion of the foresters. This cut to the heart of technocratic utilitarianism. The Forest Service did not want to have its management flexibility lessened in any way. The national forests were being managed by foresters, experts in the management of forests, and they should be allowed to make management decisions based on their knowledge, based on science. They should not be constrained by laws resulting from interest-group pressure. A wilderness bill, with statutorily designated wilderness areas, would take management decisions out of the hands of the trained experts and place them in the hands of political actors, leading to mismanagement and waste. Areas should continue to be designated wilderness by the experts, as was the case in the existing system.

A 1961 SAF survey revealed that 71 percent of members responding to a poll favored the status quo for wilderness management, and only 29 percent favored the proposed wilderness legislation. Henry Clepper, the executive secretary of the organization, testified that "in the considered opinion of a sizable majority of the professional foresters who voted in the Society's referendum of December 1961, the present procedure for establishing wilderness areas is adequate and should be continued." The results of this survey should not be surprising: as discussed above, foresters committed to technocratic utilitarianism thought that wilderness designation should be made by foresters, not through politics. Both the AFA and the SAF opposed the bills.[36]

At the first set of hearings there was another important set of groups that opposed the bill as proposed: government resource management agencies concerned that the bill would unduly limit their management authority and flexibility. The government agencies that testified against the legislation in the initial Senate hearings were the Forest Service, the National Park Service, and the California State Department of Water Resources. The Forest Service was most concerned with the threat of the bill to the administrative discretion of the agency. Chief of the Forest Service Richard McArdle commented:

> Although we are sympathetic to the general objective of the bill, we recommend that it not be enacted.
>
> . . . The bill would strike at the heart of the multiple-use policy of National Forest administration. The bill would give a degree of congressional protection to wilderness use of the national forests which is not now enjoyed by any other use. It would tend to hamper free and effective application of administrative judgment which now determines, and should continue to determine, the use or combination of uses to which a particular national-forest area is put. This balancing of conflicting demands, and the weighing of one use priority against another is the key to workable multiple-use management.

The Forest Service was specifically concerned with the potential for Congress and the National Wilderness Preservation Council to interfere with its management of the national forests; the agency feared the replacement of technical knowledge with political influence. Additionally, the Forest Service argued that these areas already had sufficient protection.[37]

The NPS opposed the wilderness bill for five reasons: the potential wilderness areas in national parks already enjoyed sufficient protection; there would be conflicts over the use of wilderness in the parks; the Wilderness Council would increase the bureaucracy between the NPS and Congress; any consideration of the wilderness bill should be delayed until a report of the nation's recreation needs was completed; and finally, the NPS had recently initiated "Mission 66," a program to develop park resources, and NPS officials feared that the wilderness bill might lead to conflicts with "Mission 66."[38] The California State Department of Water Resources opposed the bill because it feared that a wilderness system would prevent future water projects needed for irrigation, power, and general municipal supply. As discussed below, the opposition of these government agencies dissipated as the wilderness proponents modified the original proposal.[39]

The House Committee on Interior and Insular Affairs also held hearings in June 1957, which followed a pattern similar to the Senate hearings. Neither the House nor the Senate committees reported a bill following the hearings. The House would not hold hearings on a wilderness bill again until 1961 because Representative Wayne Aspinall (D, Colo.), who became chair of the House Committee on Interior and Insular Affairs in 1959, determined that the House would not consider the issue until some version of a wilderness bill had passed the Senate. Aspinall was the representative from western Colorado, a region with a large amount of public land and an area economically dependent on natural-resources industries, and he was not very sympathetic to preservationist ideas.[40]

Although these were the first of what would be many hearings on wilderness legislation, the general themes of opposition had been drawn. Throughout the next seven years, preservationists amended their proposals to lessen the opposition of various groups and interests to a wilderness system. It was true that certain groups could not be won over (namely, the commodity groups). These groups, nevertheless, could be mollified with concessions, and other groups might come to support wilderness legislation if certain changes were made.

Another issue of importance throughout the debate was primarily an institutional one: which branch would designate future wilderness areas—the executive branch or Congress? The original proposal had the executive branch making designations and Congress having a veto power. Many members of Congress wanted these powers reversed. Congress should have the primary responsibility for adding new wilderness areas by "affirmative action": Congress would pass a law, and the president could either sign the law or veto it. Such a system was more in line with standard operating procedures. When it became apparent that a wilderness bill would pass, wilderness opponents wanted expansion power in the hands of Congress rather than the executive branch because of the political realities of the period (examine, for example, testimony by the National Association of Manufacturers and the timber industry); the executive branch was viewed as being more prowilderness than Congress. Also, with Representative Aspinall chairing the House Interior Committee, opponents had an ally in a position to block further wilderness designations.

In the fall of 1957 and the winter and spring of 1958, preservationists worked to amend the wilderness legislation to lessen opposition, primarily by the Forest Service and the NPS. Senator Humphrey introduced a modified bill (S. 4028) in June 1958. The most important change gave the president the power to authorize mineral development and water development in wilderness areas if such development were determined to be in the national interest.[41]

Further Senate hearings were held on the new bill, one in Washington in July 1958, and then a series of four field hearings in the West. In the Washington hearings, the wilderness proponents gained the objective they had sought when they drafted the modified Humphrey bill. Both the Interior Department and the Agriculture Department dropped their opposition to the wilderness preservation system, though neither department became supporters of the bill. This removed a key component from the forces opposing wilderness legislation. With Forest Service and NPS opposition to the bill it was difficult to portray opponents of the bill as selfish seekers of narrow economic interests; without it, this claim could be more legitimately made.

At the western hearings, the great bulk of the witnesses who testified opposed the bill. Public opinion arriving through the mail was another issue, however. Of the mail the committee received, the great majority of which was from western states, more than 80 percent favored the bill. The most important aspect of the field hearings, though, was their timing. The November hearings virtually guaranteed that no wilderness bill would be passed in the Eighty-fifth Congress. The bill died, still in committee.[42]

New wilderness bills were quickly introduced in the Eighty-sixth Congress by Humphrey and Saylor. The Senate bill (S. 1123) contained further changes designed to soothe the still skeptical Forest Service. The bill granted the secretary of agriculture the power to control disease, fire, game overpopulation, and insects in wilderness areas. Also, the time to review "primitive" areas for reclassification was increased from ten to twenty years. Further action on the bill was delayed until early 1960 because a member of the committee, Senator Joseph O'Mahoney (R, Wyo.), had a stroke and requested that consideration of the bill be postponed until he was able to participate.

When action in executive session resumed, delays continued for a number of reasons. First, the recovered Senator O'Mahoney and Senator Gordon Allott (R, Colo.) introduced a series of amendments, and a substitute bill, that undercut the intent of the Humphrey bill. A second problem was the loss of two key wilderness supporters from the committee. Senator Richard Neuberger (D, Oreg.) died in March and Senator Murray announced his plans to retire at the end of the term, passing chairmanship of the committee to Senator Henry Jackson (D, Wash.). Finally, during the spring, Congress passed the Multiple Use and Sustained Yield Act, which statutorily reaffirmed the national forest management techniques the Forest Service had employed since the agency had gained authority over the forests in 1905. Included in the act was the phrase, "The establishment and maintenance of areas of wilderness are consistent with the purposes and provisions of this act." This clause lent further support to wilderness act opponents who claimed that new legislation was unnecessary.[43]

A new, clean prowilderness bill (S. 3809) was introduced in July that addressed some of the concerns still existing among opponents, most notably dropping the National Wilderness Preservation Council (lessening bureaucratic opposition to a new entity), removing lands on Indian reservations from potential inclusion in the system, and altering the method of expanding the wilderness system so that after fifteen years expansions could only be made by an act of Congress. Despite these modifications, wilderness legislation had spent too much time in the Senate Interior Committee, and the Eighty-sixth Congress ended without a vote on the wilderness bill.[44]

The new year, 1961, brought with it a new Congress, a newly constituted Senate Interior Committee, and a new president and his appointees. In the new Congress, a wilderness bill (S. 174), which was basically the same as the bill that had died in committee just a few months earlier, was immediately introduced by Senator Clinton Anderson (D, N.Mex.), the new chair of the Interior Committee. President John Kennedy took an active and strong role in natural resources policy. He addressed Congress on natural resources policy in February and in the speech endorsed the wilderness bill. At hearings on the bill, Secretary of Agriculture Orville Freeman and Secretary of the Interior Stewart Udall indicated their strong support for the bill. These changes may have provided the stimulus for committee action, and in July the Interior Committee reported the bill out of committee with a recommendation for passage. Despite some heated floor debate, the bill passed easily, seventy-eight to eight.[45]

With Senate passage, action shifted to the House, where the Interior Committee was more strongly opposed to wilderness legislation than the Senate Interior Committee ever had been. Field hearings took place in the West in the fall; and Washington hearings before the Subcommittee on Public Lands occurred in May 1962. These were the first comprehensive hearings in the House, and large numbers of witnesses favoring and opposing the legislation testified, essentially repeating the themes stressed in more than two thousand pages of Senate testimony. Markup of the Saylor House bill (H.R. 776, introduced in early 1961) began in the subcommittee in June. By the time the bill reached the full committee in August, all that remained of the original Saylor bill was its number. The entire bill had been replaced by a substitute bill offered as an amendment by Representative Aspinall.[46]

The bill was favorably reported out of committee in late August, and Representative Aspinall spoke to Speaker of the House John McCormack (D, Mass.) about proceeding under a suspension of the rules. Speaker McCormack balked, and Aspinall declined to go before the Rules Committee and have a rule assigned to the bill. Rather, he waited to file the report until October, with just ten days remaining in the Eighty-seventh Congress. The bill did not receive a rule and never came up for a vote on the floor. Wilderness legislation failed again and would have to begin anew in both chambers in 1963.[47]

Following the pattern of the previous six years, the Senate was the first body to act. Senator Anderson introduced a bill (S. 4) in January that was identical to his bill that had passed the Senate in 1961, hearings were held in February and March, the bill was favorably reported out of committee in April, and it was passed seventy-three to twelve on April 8. In the House, twenty-four wilderness bills were quickly introduced. Hearings were held in the West and in Washington, with the same themes and perspectives reiterated.[48]

During this period, wilderness advocates received further evidence to bolster their position. In 1963, the Forest Service reclassified the 1,875,306-acre Selway-Bitterroot Primitive Area located in Idaho and Montana. Due to intense pressure from timber interests, only 1,239,840 acres were declared wilderness; the remaining acres were made available for multiple-use management. This was yet another example of the Forest Service's tendency to create "starfish wilderness" areas (central mountain peaks with arms of rocky ridges, lacking forested valleys), wilderness advocates claimed.[49]

In the House, it quickly became apparent that attention would focus on three bills: Anderson's bill that had passed the Senate (H.R. 930), a version of the Anderson bill modified somewhat to meet the concerns of the opposition that was introduced by Representative Saylor (H.R. 9070), and a bill introduced by Representative John Dingell (D, Mich.), which went even further than the Saylor bill in placating the wilderness opponents (H.R. 9162). In 1963, President Kennedy met with Representative Aspinall to discuss wilderness legislation, and both sides agreed to try and work out an acceptable compromise. Most interests, both in favor of and opposed to a wilderness system, thought the Dingell bill to be the most likely candidate to emerge from the Interior Committee. Even that bill appeared in jeopardy in April, however, when Aspinall reacted angrily to a *Washington Post* editorial that blamed him for blocking wilderness legislation. Just a few days later Howard Zahniser, the chief wilderness advocate, died of a heart attack. This appeared to soften Aspinall, and the Subcommittee on Public Lands began to mark up a bill in June.[50]

Surprisingly, the subcommittee chose to mark up the Saylor bill rather than the Dingell bill. The subcommittee adopted one major amendment, one that would allow mineral exploration and exploitation in wilderness areas for twenty-five years. In the full committee, an amendment was adopted to delete a specific area in California, the San Gorgonio Wild Area, from wilderness classification to allow for the development of a ski area there. The bill was reported to the full House in June, and during floor debate an amendment was passed restoring the San Gorgonio area to the wilderness system. The House approved the bill 373 to 1 in July. A conference committee was formed, and it largely followed the House or Saylor bill, with the exception that the mining exemption was reduced to nineteen years. The conference report was adopted by both House and Senate on August 20, and it was signed into law by President Johnson on September 4, 1964.[51]

As finally passed, the Wilderness Act was a compromise among preservationists and those opposed to the wilderness idea and system. The preservationists were able to achieve a statutorily designated wilderness system, replacing the more tenuous administrative system. The act included a number of concessions. Although eventually wilderness areas would be closed to new mining claims, the lands were to remain open to mining entry until December 31, 1983, and the development of valid claims would be allowed indefinitely. The wilderness canoe area in northern Minnesota was to be managed in a manner inconsistent with the Wilderness Act, with motorboat use and some timber harvesting to continue. Established grazing practices in wilderness areas were allowed to continue, though no future grazing expansions would be allowed. And as reported above, along the way preservationists had to drop the proposed National Wilderness Preservation Council, drop Indian lands from the system, give the president the power to authorize water developments within wilderness areas, and give the secretary of agriculture the power to use measures to control fire, insects, and disease, and approve special nonconforming uses in wilderness areas.

The act designated 9.14 million acres of wilderness in fifty-four areas, all on national forests. The passage of the act, however, did not mean that the wilderness battle was over. From 1964 through the present, battles have raged over the expansion of the wilderness system and the specific meaning of the act in Congress, in the courts, and in the executive branch. Most notably, the Roadless Area Review and Evaluations (RARE I and II) undertaken by the Forest Service have been vast programs of public participation, information processing, and political compromise designed to determine which national forest lands were suitable for wilderness classification. As of 1995, 35 million acres of Forest Service lands were designated as wilderness, and the process of designating more lands as wilderness continues.

Despite compromises from the original bill to allow for potential mining and water development and continued grazing, the Wilderness Act represented the most important change in national forest policy since the formation of the Forest Service. Although less than a tenth of national forest land was affected, it was the first successful challenge to the privilege of technocratic utilitarianism within the Forest Service. The reduction in management discretion and flexibility of the agency was not only important in terms of wilderness management; it also delivered the first blow to the institutionalization of privilege for technocratic utilitarianism. Indeed, the further erosion of expert authority in the management of the national forests is demonstrated in the clear cutting controversy of the 1960s and 1970s in which the public openly questioned timber harvesting techniques. Although clear cutting has continued, public participation in Forest Service decision making has continued to increase as well. The formation of the Association of Forest Service Employees for Environmental Ethics (AFSEEE), a group within the agency strongly influenced by preservationism, is further evidence of the lessening privilege of technocratic utilitarianism.[52]

How can we explain the success of the supporters of preservationism in the face of the privilege of technocratic utilitarianism? We must turn to an examination of the state for the answer. The competition between the Forest Service and the NPS over the management of recreational and scenic lands was perhaps the chief impetus for the creation of the administrative wilderness system in the Forest Service. Without such a system, Forest Service officials feared that they would lose further portions of the national forests to the NPS for new or expanded national parks. Once the administrative wilderness system was in place, thanks to preservationist mavericks such as Leopold, Carhart, and Marshall, the idea of a wilderness system was less foreign to the technocratic utilitarians who dominated the ranks of the Forest Service. Although opposition within the Forest Service to the idea of wilderness remained, it was certainly less than it would have been if the wilderness advocates were trying to introduce an entirely new program in their proposed wilderness legislation. Hence, when the wilderness legislation was working its way through Congress, Forest Service opposition was based primarily on political interference with administrative discretion in the management of the forests, rather than opposition to the entire program.

The shape of the state did not work only in favor of preservationists. The institutionalization of extralegal property-rights systems in the hard-rock mining and grazing policy regimes also helped to shape the outcome of the wilderness issue. Based on the Mining Law of 1872, miners asserted a property right on any public land they thought might have minerals. Ranchers turned to the Taylor Grazing Act to justify their property rights to public lands they used for grazing. Both of these uses based on existing property rights were allowed to continue.

This case demonstrates the complexity of the interplay of ideas, groups, and the state. Although technocratic utilitarianism had been firmly embedded in the state from the inception of the forestry policy regime, this privileged idea was wounded by the Wilderness Act. The Forest Service had acted to protect its turf and had created an administrative wilderness system. This provided the leverage necessary for the supporters of preservationism to gain their goal. Of similar importance, the wilderness issue was *the* issue for preservationists; it was one of many for the technocratic utilitarians. Once the act was passed, though, the challenge to technocratic utilitarianism was not confined only to wilderness; it spread. Although the idea remained firmly embedded, the professional policy pattern of the forestry regime began to unravel in the late 1960s.

The Privatization Initiative of the 1980s: Increasing Economic Efficiency through Private Management?

Ever since the federal government began reserving or withdrawing lands and making them unavailable for disposal, there have been groups and individuals who have sought to have these policies overturned. The rationale of their arguments has been based on combinations of personal gain and a particular conception of the public interest: that the public benefits the most when property is owned and managed by private individuals—economic liberalism. Since the beginning of government retention in the 1890s, movements to transfer large portions of the public lands to either the states or the private sector have arisen. Significant movements in the 1940s and 1950s concentrated on the transfer of Grazing Service and BLM lands. In the 1970s another protransfer movement developed in the western states, the Sagebrush Rebellion.[53]

The flames of the Sagebrush Rebellion began when Nevada and Arizona requested that the federal government grant further public lands to the states. These flames were stoked by the passage of the Federal Land Policy and Management Act (FLPMA) in 1976, which further alienated western states. There were two main causes of the rebellion. First, Sagebrush rebels argued that federal ownership and management had an adverse effect on the economies of western states. If the lands were in state or private hands, they would be used to increase economic development in the West. And second, they argued that the states' should control this vast land within their borders because federal control undermined the sovereignty of these states.

The states used a tripartite strategy in their efforts to have the lands transferred: they used the courts to claim legal rights to the lands under the United States Constitution, they passed laws in state legislatures claiming the lands, and their representatives in Congress introduced legislation to transfer the lands. None of these strategies proved successful. Both Nevada and Utah lost cases in the federal judicial system based on Constitutional arguments. Five states—Arizona, Nevada, New Mexico, Utah, and Wyoming—passed acts claiming title to the public-domain lands within their states, but none pressed their cases. And in Congress, sixty bills dealing with the Sagebrush Rebellion were introduced in the Ninety-sixth Congress (1978–80), but no specific hearings were held and no legislation made it out of committee.[54]

The Sagebrush Rebellion was defused in 1980 with the election of Ronald Reagan, who favored states' rights and reduced government regulation. His first secretary of the interior, James Watt, implemented a "good neighbor" policy that emphasized better national-state relationships on public-lands issues. The success of this program, combined with the poor legal arguments and the lack of a political consensus in the West favoring the program, removed much of the fuel from the flames of the Sagebrush Rebellion.

Not long after the rebellion was defused, however, a new transfer movement with different, though related, roots arose. This movement advocated transferring land from the federal government to the private sector. The disposal of some public lands was part, albeit an important part, of a larger privatization movement designed to move numerous federal properties and services into the private sector. In this movement, we see the idea of economic liberalism still very much alive in public-lands policy.

This new privatization movement was essentially an intellectual movement born and nurtured by an informal grouping of western economists who proposed a "New Resource Economics" (NRE). The basic theme of the NRE is that the private market is best suited to determine the allocation of resources in society; in other words, economic liberalism. The privatization movement was essentially a property-rights movement. Advocates of privatization argued that when land and resources are held publicly, there is an imperfect assignment of property rights and such resources are not allocated efficiently. In order to overcome this problem, we should privatize the public lands, because "when rights are both privately held and easily transferable, decision makers have easy access to information through bid and asked prices, as well as an incentive to move resources to higher valued uses."[55]

Proponents of the NRE argue for privatizing public lands used for a variety of different primary uses, from wilderness to grazing to forestry. In focusing on the national forests, economists Richard Stroup and John Baden argue forcefully that "the U.S. Forest Service systematically supports inefficient timber production." They continue by arguing for a change in ownership patterns: "Privatizing the national forests should end many of the obstacles to good management. Not only would decision makers be given larger amounts of validated and continuously updated information, but political obstacles to efficient management would largely disappear." Also arguing for an end to "a failed experiment in socialism," forestry professor Barney Dowdle states: "Government should be removed from the business of producing timber. The free enterprise system has demonstrated convincingly that it can handle this task much more efficiently. Privatization of commercial government timberlands would relieve society of an unnecessary and costly tax burden, the magnitude of which can only increase. Perhaps more important, all Americans would benefit from the expanded freedom provided by the restoration of private property rights and the free enterprise system."[56]

A key figure in translating NRE theory into government policy was Steve Hanke, a senior economist on the president's Council of Economic Advisors in the early 1980s. In numerous articles outlining and advocating the program, Hanke noted the benefits that would result from privatization. The productivity of the lands transferred to the private sector would increase because they would be managed more efficiently, and the remaining public lands would also be managed more efficiently (because scattered and isolated parcels would have been privatized). Relatedly, transferring the lands to the private sector would make the lands more accountable to the market, and hence, more respective of consumer preferences. The sale of these lands would generate revenues that could be used to reduce the national debt. Negative cash flows would be reduced or eliminated, because the federal government often spends more on resource management on the public lands than it receives in revenues from the lands (in fiscal year 1979, for example, the Forest Service spent $1,595 million and received receipts of $780 million, resulting in a negative cash flow of $815 million). Land-use decisions would become less politicized because the lands would be moved from the public to the private sector. And finally, transfer of these lands to the private sector would increase the size of the state and local tax base. Hanke concludes one article in a more dramatic manner: "The real issue in the privatization debate is coming to the forefront: the choice between capitalism and socialism, or between private property and individual freedom versus public ownership and serfdom."[57]

David Stockman, director of OMB, was also a major supporter of the privatization initiative. From his perspective, four types of land were prime candidates for disposal: lands that could not be effectively managed due to the small size or isolation of the parcels, lands near communities that could contribute to local economic development, properties managed by the Corps of Engineers and the Bureau of Reclamation that were not really necessary to achieve public purposes, and properties that have potential for higher and better use in private ownership, which could include almost any parcel.[58]

The Asset Management Program (AMP) was essentially born at a Cabinet Council on Economic Affairs meeting in February 1982. Donald Regan, chair of the council, sent a memorandum to President Reagan with three recommendations:

(1) That a Presidential Property Review Board be established by Executive Order to oversee the property review process.

(2) That the Cabinet Council Working Group on the Sale of Federal Property develop legislation to place all proceeds from federal property sales in a fund or account for use only to reduce the national debt.

(3) That the Cabinet Council Working Group on the Sale of Federal Property coordinate a comprehensive review by affected federal agencies of existing statutes and regulations pertaining to sales of federal lands and develop appropriate legislative proposals to expedite such sales in a cost-effective manner.

All three recommendations were approved by President Reagan.[59]

In making these recommendations, the council was aware of the political controversy any program to sell significant portions of the public lands might generate. Among the segments of society that would be upset would be ranchers, environmental groups, local communities in the West, and citizens who used the lands for recreational purposes. Nevertheless, "the Working Group . . . recommends promptly developing a program to dispose of unneeded public lands."[60]

From the beginning of the movement to retain some public lands in the 1890s, the technocratic utilitarians had argued that the market did not adequately function in the forestry realm, and this perspective had not changed. Proper forest management involved considerations in addition to profit and economic efficiency, such as the multiple uses of recreation, watershed protection, wilderness, and wildlife habitat. Both the AFA and SAF opposed the privatization initiative. For preservationists, the public lands were the heart and soul of what they viewed to be an important, central part of a good society: lands preserved from exploitation for the benefit of the present and future generations. Preservationists viewed privatization as a major threat to this heart and soul, and to the successes they had achieved over the last eighty years. Thus, it was the preservationists who led the attack on the privatization movement. By this time, the preservationists had also been successful in convincing the majority of the population that some amount of public-lands preservation was in the public interest, and the preservationists drew upon this support in their successful effort to stymie the privatization movement.

In response to the cabinet council recommendations, President Reagan issued Executive Order 12,348 on February 25, 1982. The executive order established the Property Review Board, which included the counselor to the president, the director of OMB, the chair of the Council of Economic Advisors, the assistant to the president for policy development, the chief of staff and assistant to the president, the assistant to the president for national security affairs, and other persons the president may choose to appoint. (Interestingly, no representatives of the Agriculture or the Interior Departments were assigned to the board.)

The executive order specified a number of functions for the newly created board, chiefly to review existing acquisition, utilization, and disposal policies and to develop "standards and procedures for executive agencies that are necessary to ensure that real property holdings no longer essential to their activities and responsibilities are promptly identified and released for appropriate disposition." In order to help achieve the goals of moving unnecessary lands out of the public sector, the heads of the land-holding federal agencies were directed to report to the Property Review

Board the use status of current lands and to identify lands that were no longer necessary for the achievement of federal purposes. The fiscal year 1983 budget proposal called for the sale of 5 percent of the nation's land (excluding Alaska), approximately thirty-five million acres, over five years. The revenues projected under the program were $1 billion for fiscal year 1983, and $4 billion per year for fiscal year 1984 through fiscal year 1987, for a total revenue of $17 billion.[61]

Most of the public response to the AMP was negative, some was ambiguous, and very little was supportive. The main source of opposition came from the environmental community. Virtually all of the major environmental groups strongly opposed the AMP, including Friends of the Earth, the National Audubon Council, the National Wildlife Federation, the Public Lands Institute, the Sierra Club, the Trust for Public Lands, and the Wilderness Society. These groups made their opposition known at congressional hearings, in the media, and through their own publications.

At the first congressional hearings on the AMP initiative, a spokesperson for the Wilderness Society argued:

> It is time for the privatization scheme to be revealed for what it is, a land grab to provide immense profit to a few at the expense of present and future generations. . . .
>
> If the public land sale program is not carefully kept in check by the Congress, we fear that historians of the future will judge us harshly indeed. "Privatization" could become known as *piratization* of our Nation's public land and resource inheritance. The 1980s could well be viewed as another era of corruption and scandal, in which these lands were put on the auction block, sold to the highest bidder and the future be damned.

At the same hearings, a representative of the Sierra Club focused on the economic rationale of the issue:

> The underlying premise of "privatization" is one with which we strongly disagree, and which runs counter to the public land legislation of the last few decades. That premise, as it is reflected to varying degrees in all of the proposals, is that public property should be managed for its highest economic return.
>
> Economic return cannot be used as the sole yardstick for measuring public benefit from federally owned property. Public benefit must be computed using a more complicated formula which considers other values, for what serves the public interest does not always provide the highest economic return.[62]

Another segment of opposition was the groups most associated with forestry and technocratic utilitarianism, namely, the AFA and the SAF. Both groups opposed the AMP generally, and specifically the proposed legislation to give the Forest Service broader authority to dispose of national forest lands. In House hearings, the AFA offered testimony opposing the AMP:

> We hope the Department of Agriculture is not planning to high-grade the national forests on a grand scale and open the door to a going-out-of-business sale for the Forest Service.
>
> ... We strenuously oppose the logic that the multiple land use system should be sold or otherwise conveyed to the private sector as a means for raising revenue or converting such lands to higher economic purposes, or merely as a means of dissolving the Federal estate as a cost-cutting device.

At later House hearings, the SAF stressed the need to manage public lands for reasons other than economic efficiency: "We view with much skepticism the development and implementation of the President's asset management program. Early projections of multimillion dollar revenues associated with a massive program of surplus land disposal raised great concern with most resource-management professionals. . . . However, we do not accept the profit motive, the raising of revenues to assist in retiring the national debt, as the primary objective of any land disposal program."[63]

Many state governments also reacted negatively to the AMP. The idea of selling scarce public lands in the East, which are heavily relied upon for recreation, met with strong criticism. Numerous eastern states passed resolutions opposing the sale of any public lands within their states, including Missouri and Ohio. Additionally, members of Congress from states in which sales were proposed and were subsequently strongly opposed by the public—including Arkansas (Senator Dale Bumpers, D), Minnesota (Representative Bruce Vento, D), Ohio (Representative John Seiberling, D), and Wisconsin (Representative Les Aspin, D)—testified against the AMP throughout the course of the hearings.

Even in the West criticism was strong. Governors Bruce Babbitt of Arizona, Ed Herschler of Wyoming, Richard Lamm of Colorado, and Scott Matheson of Utah all testified in opposition to the privatization initiative. Governor Lamm made his opposition clear before the Senate Committee on Energy and Natural Resources:

> If the public lands are returned to private ownership under these or any other proposals, this Nation and the generations of Americans yet to

follow will suffer a loss many times greater than the one-time fiscal harvest could ever hope to provide. . . . I believe that the disposal of public lands and resources under any disguise, be it through a property review board or a loose interpretation of well-intended resolutions, ignores the public interest, congressional intent, and sound land management principles.

The California secretary for resources expressed that state's opposition to the AMP: "The so-called privatization movement program of the current administration is a fantasy based on a cornucopian philosophy of unlimited resources more appropriate to the 19th century than to the 20th. The nonsense of attempting to retreat and turn back 50 years in our resources use approach is to ignore the stress of resource consumption and the need to manage rather than exploit."[64] As was the case with their eastern colleagues, numerous western members of Congress testified in opposition to the AMP, including Senator Henry Jackson (D, Wash.), Representative Ron Marlenee (R, Mont.), and Representative Jim Santini (D, Nev.). Even Nevada state senator Norman Glazer, an initiator of the Sagebrush Rebellion, testified against the initiative. Sagebrush rebels wanted to see less land under federal control, but they felt that this program had too many negatives.

Among commodity user groups, the response to the AMP was ambiguous. Some groups and operations favored privatization, others did not. The reason for the opposition was the fear that smaller operations would not be able to buy property they had been using in a free-market setting because they did not have the necessary capital. Also, many commodity users were subsidized by the federal government via the low fees they had to pay to use resources. In the grazing realm, privatization could disrupt the complex system of public-private grazing rights that had developed in the West. The executive secretary of the Nevada Cattlemen's Association claimed that "cattlemen would have supported the sale of public lands if there had been provisions built into the proposals to ensure that they would be able to utilize their present rights." At the national level, grazing interests supported the general idea of privatization, but with some important reservations concerning the existing property rights ranchers claimed on these lands. The Burlington Northern Timberlands operation was upset with the moratorium on private-public land exchanges during the AMP development phase. Rather than favoring the opportunity to buy government land, the corporation argued that "all land exchange possibilities should be exhausted before federal squares in the pattern are offered to third parties."[65]

The National Association of Counties also greeted the program with some skepticism. The group looked "favorably on the general concept" of the AMP, but had two main reservations concerning the program: they opposed the end of the free or discounted transfer of federal lands to state and local governments, and they were concerned with the lack of local consultation on the program. Unless these considerations were addressed, they would not support the initiative. Even conservative Republican senator James McClure of Idaho, chair of the Energy and Natural Resources Committee, was skeptical of the program.[66]

At the three congressional hearings devoted to the AMP (two in the House, one in the Senate), only two witnesses clearly supported the program: Representative Ken Kramer (R) of Colorado and the National Association of Realtors. Kramer was convinced that the AMP could eliminate the entire federal deficit. The rationale for the realtors support was clear: realtors would benefit a great deal under a program that transferred significant amounts of public lands to private ownership. In their congressional testimony, however, they presented a more philosophical rationale for supporting the AMP: "Real property in the private sector automatically is drawn to its best use by the continuous valuation of property by the public. The Government does not have such an automatic system of adjustment but tends to hold real property often to the detriment of the people supposedly represented by Government."[67]

Amidst the public debate over the AMP, the administration continued to develop the AMP initiative. After working out the general framework of privatization and making broad proposals in 1981 and early 1982, more specific figures on acreage to be sold and on criteria for disposal began to appear in mid-1982. The first public information on the sale of specific Forest Service lands came at an oversight hearing of the House Subcommittee on Public Lands and National Parks of the Committee on Interior and Insular Affairs in June 1982. Deputy Secretary of Agriculture Richard Lyng testified that both the Agriculture Department and the Interior Department were instructed to develop programs in which each department could dispose of $2 billion worth of property in fiscal year 1984. At the time of the hearings, the Forest Service had identified 833 acres for disposal, primarily administrative sites, under the authority of the Federal Property and Administrative Services Act. Lyng also testified as to the acreage to be made available for disposal in the Forest Service preliminary inventory: 353,700 acres of fee-simple land, 1,575,300 acres of split mineral-rights land, and 1,0445,400 acres of land in which the Forest Service held an interest via public purpose reverter clauses. Lyng concluded by making it clear that the Forest Service planned to seek increased land disposal authority.[68]

Two months later, Secretary of Agriculture John Block announced that the Forest Service would seek congressional authority to allow for the general disposal of land the following year. He also announced that the approximately 190 million acres of land under the control of the Forest Service had been classified into three groupings in terms of land sales. First, 51 million acres of land would be retained: wilderness areas, recreation areas, wild and scenic rivers. Second, 60,133 acres of land were to be sold, primarily Land Utilization plots acquired in the 1930s that the Forest Service did not need congressional authorization to sell. And third, the remaining land, close to 140 million acres, was classified for further study. From this pool of lands, 15 to 18 million acres were to be identified and studied extensively for potential sale. According to Block, the main focus would be upon scattered holdings, lands leased for private use, and areas within national forests and national grasslands in which there was extensive private ownership. The response by environmental groups was quick and negative. A representative of the Wilderness Society stated that "Block's initiative is another step in the administration's campaign to turn over the wealth of the public lands to corporate interests at giveaway prices."[69]

Assistant Secretary of the Interior Garrey Carruthers and John Crowell, assistant secretary for natural resources and the environment of the Department of Agriculture, further elaborated on the AMP in a September 1982 hearing. Crowell stated that 60,000 acres of Forest Service land had been identified for sale, land the Forest Service did not need authorization to sell. He also specified the categories of the 18 million acres to be studied for disposal: (1) 300,000 acres that would improve overall administration, (2) 20,000 acres of special-use permit areas, (3) 5 million acres to be made available after modifying the exterior boundaries of national forests, and (4) 12.6 million acres from portions of national grasslands and other units composed of two or more separated areas where receipts are less than resource management and protection costs. According to Crowell, work on developing the necessary land-disposal legislation was just beginning. "We view this legislation," Crowell testified, "as an opportunity to improve the National Forest System by disposing of land that cannot be managed efficiently; by disposing of land that will serve community development by being in private ownership; and by disposing of land with potential for higher and better use in private ownership or under management of other units of government."[70]

By early 1983, land sale goals throughout the administration were already being reduced by 70 percent, a harbinger of the eventual fate of the entire AMP. In 1982, fiscal year 1984 projections were for $4 billion in revenue from the AMP, $2 billion from the GSA, and $2 billion from Agriculture and Interior. The revised figures projected fiscal year 1984 sales of $1.15 billion: $650 million from GSA, $300 million from Interior, and $200 million from Agriculture.[71]

In March 1983, the Department of Agriculture announced that it planned to seek legislative authority to dispose of up to six million acres of land managed by the Forest Service (national forests, national grasslands, and land-utilization projects), approximately 3.2 percent of the lands in the system (3.6 percent of the lands excluding Alaska). These six million acres were selected because the parcels were isolated, were in checkerboard ownership, were single-purpose lands, or were needed for community expansion. The lands proposed for disposal were located in thirty-nine states throughout the country, more than one-third of the total acreage coming from Montana, California, and Colorado (see table 3). After obtaining this legislative authority, the Forest Service would then conduct intensive studies of the previously identified six million acres and determine which specific parcels were suitable for disposal. The Forest Service hoped to have a legislative proposal ready to submit to Congress within one or two months.[72]

Opposition to the Forest Service proposal was quick and predictable. Gaylord Nelson, counsel of the Wilderness Society, described the proposal as an "opening salvo in an unprecedented assault" on the national forest system. Another Wilderness Society spokesperson elaborated: "Two-thirds of the timber in this country comes from private land, and private timber industries can overcut or do whatever they want with that land and the timber on it. But we want to make sure that the other third is managed properly so that we will all have timber into the 21st century." Hearings on proposed BLM land sales in Idaho just a few days later demonstrated that opposition to the AMP was widespread in the West. Among those opposing the privatization of public lands in Idaho were nine western governors, environmental groups, hunters, fishers, and ranchers.[73]

In July 1983, Secretary of the Interior Watt removed Interior Department lands from the AMP. Watt viewed the program as a political mistake and a liability to President Reagan because it was undermining the political good will created through the good neighbor program. Even conservatives such as Senator Jesse Helms (R, N.C.) and Senator McClure opposed the program. Watt's action did not effect the Forest Service program, but the momentum was clearly gone from that program as well. Five months had

Table 3. Proposed Forest Service Parcels for Disposal, by State

	Number of Acres	Amount of National Forest Land in the State (%)
Alabama	57,068	9
Alaska	—	—
Arizona	133,982	1
Arkansas	47,089	2
California	723,975	4
Colorado	442,323	3
Connecticut	—	—
Florida	5,297	.5
Georgia	130,150	15
Hawaii	—	—
Idaho	186,709	.9
Illinois	69,694	27
Indiana	13,318	7
Kansas	15,720	15
Kentucky	34,593	5
Louisiana	66,090	11
Maine	—	—
Michigan	239,139	9
Minnesota	237,144	8
Mississippi	271,081	24
Missouri	165,367	11
Montana	872,054	5
Nebraska	24,132	7
Nevada	18,004	.3
New Hampshire	2,370	.3
New Mexico	269,952	3
New York	—	—
North Carolina	71,358	6
North Dakota	239,239	22
Ohio	63,093	36
Oklahoma	61,088	21
Oregon	277,091	2
Pennsylvania	15,154	3
South Carolina	23,322	4
South Dakota	185,332	9
Tennessee	6,523	1
Texas	158,905	20
Utah	141,675	2
Vermont	19,729	7
Virginia	48,577	3

Table 3. *Continued*

	Number of Acres	Amount of National Forest Land in the State (%)
Washington	317,808	5
West Virginia	42,011	4
Wisconsin	41,541	3
Wyoming	334,295	4
Puerto Rico	—	—
Virgin Islands	—	—
Total	6,071,993	3.2

Source: "Agency Weighing Sale of 3.2% of Forest System," p. 16; "6 Million Acres of U.S. Forest Eyed for Sale," p. A8.

elapsed since the program had been announced, and still no legislative proposal for disposal authorization had been sent to Congress.[74]

The figures for fiscal year 1983 revealed that the first year of the land-sales program was quite disappointing. The administration had established a goal of $1.23 billion in revenue generated by the sales, but the actual revenue was approximately one-tenth of that, less than $198 million ($194.7 million from GSA sales and $3 million from BLM sales). The administration blamed the low sales on a soft real estate market, turf-protecting agencies, and political opposition.[75]

In November 1983, Congress reacted legislatively to the AMP. In the Appropriations Act for the Interior Department and Related Agencies for fiscal year 1984, Congress expressed its skepticism of the AMP:

The Congress finds that the Forest Service's proposal of March 15, 1983, to consider six million acres of the national forests for possible sale has met with considerable opposition; and the national forests are an important part of the national heritage of the United States; and the national forests provide and protect important resources; and the national forests provide unique opportunities for recreation; and it is inconsistent with past management practices to dispose of large portions of our national forests. It is, therefore, the sense of the Congress that it is not in the national interest to grant the authority to sell significant acreage of the national forest until such time as the Forest Service specifically identifies the tracts which are no longer needed by the Federal Government; inventories the tracts as to their public benefit value; provides opportunities for public review and discussion of the tracts; and completes all necessary environmental assessments of such sales.

This legislation appeared to put another shovel full of soil onto the casket of the AMP.[76]

In a February 1984 oversight hearing, the Forest Service seemed to accept the death of the AMP. The following exchange between Seiberling, the subcommittee chair, and Dale Robertson, associate chief of the Forest Service, about the national forest land-disposal legislation is illustrative:

R: That piece of legislation has been delayed and there is no target date.
S: Permanently?
R: I wouldn't say permanently, but there's no sense of urgency any longer.
S: At least—probably till after November 6th anyway.
R: We're not actively working on that right now.[77]

In the end, the disposal never occurred. Even with an administration in office that favored economic liberalism, those supporting privatization could not overcome advocates of technocratic utilitarianism and preservationism. The fragmentation of land-management responsibilities that had developed made it quite difficult to develop a coherent and coordinated privatization program. The administration attempted to address this problem with the creation of the Property Review Board, but the board could not overcome these problems in its efforts to direct the policy. This fragmentation is further reflected in the legal regime that had developed regarding the public lands beginning in the 1780s. This complex legal regime is apparent in the different disposal authorities, distinctions between public-domain land and lands acquired from private ownership for specific purposes, and distinctions between lands reserved for specific purposes and unreserved lands. This fragmentation certainly made it more difficult to dislodge an existing privileged idea regarding forest lands.

The fundamental difficulty of the privatization challenge, though, was that it contradicted the idea of the public lands. In the 1890s supporters of technocratic utilitarianism and preservationism had successfully joined forces in their efforts to convince the government to retain significant amounts of public lands in government ownership. In the 1980s the coalition that helped found the national forest system reformed to block the privatization movement. By this time, technocratic utilitarianism had been institutionalized in the forestry policy regime for eighty years, yet the privilege of this idea had lessened since the 1960s. Preservationism was on the upswing. It is difficult to determine if the weakened privilege of technocratic utilitarianism would have been able to hold off the supporters of privatization without the help of the preservationists, but it is clear that the coalition of supporters of both ideas was of great importance in this policy debate.

Conclusion

The existence of the public lands owes much to the movement to reserve forests on the public lands in the 1890s. This movement was successful in altering the dominant land management philosophy at the time, economic liberalism. The ideas of technocratic utilitarianism and preservationism, both of which held that it was in the public interest for these forest lands to remain in public ownership, were the basis of a successful coalition in the 1890s and the early 1900s. It was technocratic utilitarianism, though, that became embedded within the state in the forestry policy regime. This idea was pervasive in the forestry profession and among the foresters who would manage these forests. Embedded at the creation of the national forest system, this idea served to shape a professionalized policy pattern in the forestry regime that was virtually unchallenged through the 1960s.

The Wilderness Act was a severe jolt to the privilege of technocratic utilitarianism. The law, representing a central goal of preservationism, allowed nonforesters to dictate to foresters how they would manage certain lands. This crack in the foundation of privilege for technocratic utilitarianism spread to issues of forest management and public participation. A further challenge, the privatization initiative, was launched by supporters of economic liberalism in the 1980s. This challenge failed as the privilege of technocratic utilitarianism within the state, together with the support of societal groups supporting preservationism, was able to stop the initiative.

Although technocratic utilitarianism remains privileged in the state, a role it has held since the inception of the forestry policy regime, this privilege has lessened considerably since the late 1960s. The professional policy pattern that characterized this regime for so many years is in the midst of collapse in the face of the increasing influence of preservationism, economic liberalism, and interest-group liberalism, within both the state and society. If current trends continue, the remaining privilege of technocratic utilitarianism will soon disappear, to be replaced by a more open and pluralistic policy regime.

MANAGING
THE NATION'S
GRAZING LANDS
BEEF, WOOL, GRASS,
AND POLITICS

A federal grazing policy regime was not established until the 1930s, later than the other regimes examined. In this case, interest-group liberalism was embedded within the state with the passage of the Taylor Grazing Act. The act, designed to give ranchers control of grazing policy, initiated a captured policy pattern. Interest-group liberalism was challenged by supporters of technocratic utilitarianism and preservationism in the debate over the passage of the Federal Land Policy and Management Act. Supporters of each idea came away from the law with some victories, and the privilege of interest-group liberalism began to tarnish. The final case, the debate over grazing fees in the 1970s and 1980s, demonstrates the continued weakening of the privilege of interest-group liberalism. Like the previous cases examined, in the grazing policy regime an idea was embedded and shaped policy through the 1960s, when it became the target of increasing challenges from supporters of other ideas. In the 1990s, the embedded idea is still privileged within the state, but the policy pattern it helped to shape is dissolving.

The Taylor Grazing Act: Resource Management for Public-Domain Grazing Lands

In the late 1800s, grazing on the public lands was widespread and unregulated. As was the case during the mineral rushes of the 1850s and 1860s, the federal government had not established a policy for grazing on these public lands, but sheep and cattle ranchers grazed on the lands nonetheless. In some areas, as the miners did, these ranchers also established extralegal property-rights systems to fill the void left by the government and to help reduce anarchy on the range.[1]

The federal government first became involved in grazing policy with the creation of the forest reserves. At first, grazing on the reserves was prohibited. But the policy was changed in 1898 to allow grazing, and by 1900 all grazing on the reserves was based on a permit system under the control of the secretary of the interior. Following the transfer of the administration of the reserves to the Department of Agriculture, the Forest Service began to charge fees for the grazing permits. Ranchers were strongly opposed; they even challenged the constitutionality of these fees (the Supreme Court ruled them constitutional). Despite their opposition to the fees, stockmen in general supported the permit system because it added stability and order to the sometimes chaotic public-lands grazing.[2]

This support for order is demonstrated in cattlemen's support to have a leasing system established on the public domain lands. In 1901 a bill was introduced to establish such a system. Despite the backing from cattle interests, the Public Lands Commission, and President Roosevelt, this initiative was blocked by the opposition of settlers and farmers, who did not want to see land tied up in ranching, and sheepmen, who relied on a more transient grazing approach. Each year through 1916 the leasing proposal met with defeat.[3]

Following the passage of leasing programs for fossil fuels and water development in 1920, it appeared that a leasing program for grazing would soon be passed as well. Passage, however, was delayed until 1934. The grazing policy contained in the Taylor Grazing Act was a policy primarily shaped by specific interests and the ideas of economic liberalism and interest-group liberalism, an idea descended from economic liberalism.

The stockmen who played such an important role in shaping and passing the legislation did not present a united front. Rather, both variants of liberalism were represented: some stockmen favored transferring the public-domain grazing lands to the private sector (namely, to them) to achieve the benefits of private property. Some advocated transferring the lands to the states as either an intermediate step on the path to private ownership or an alternative superior to federal ownership and control (because the states would be more sympathetic to the economics of the livestock industry and would be more susceptible to political pressures applied by stockmen). And some wanted the federal government to administer a leasing system controlled by ranchers (interest-group liberalism). This approach was favored for its political pragmatism and due to the cost of purchasing the necessary land. All were united in the goal of bringing stability to the livestock industry and the range. The approach a rancher supported was based on some combination of the rancher's view of his interests and his idea of how society should work.[4]

Supporters of technocratic utilitarianism and preservationism were not very important in the formation of grazing policy on the public-domain lands. The profession of range management was, at best, just beginning (the Society of Range Management did not form until 1948). Hence, there was no trained cadre of range managers advocating the scientific management of public-domain grazing lands. Preservationists were not interested in the issue; their energies were focused on the preservation of scenic areas, recreational areas, and spiritual areas, which at this time did not include the often stark and arid grazing lands.

Although Congress actively considered the issue from the mid-1920s, a law was slow in coming. In addition to the disarray among stockmen, a battle between the Agriculture Department and the Interior Department over the administration of the program slowed action. A bill introduced by Oregon senator R. N. Stanfield (R) to establish a leasing system managed to make it out of committee in 1925 but was defeated on the floor. Additional bills to establish a grazing-lease system on public-domain lands were introduced each year through 1929, but with no success. The entire issue was displaced from 1929 through 1932 by President Herbert Hoover's unsuccessful proposal to transfer public-domain lands to the states.[5]

In 1932, the Hoover administration drafted and introduced a bill for the creation of grazing districts on public-domain lands based upon the study of past bills and discussions with Agriculture and Interior and representatives of grazing interests in Congress. The bill (known as the Colton bill after its chief sponsor, Utah representative Don Colton [R]) passed the

House but was still in committee in the Senate when the Seventy-second Congress came to an end.[6]

The next year, attention focused on legislation introduced by Representative Edward Taylor of Colorado (D). Debate on the bill followed the usual pattern of opposition and support, with the exception that this time the bill passed both chambers and became law. Opponents concentrated, as in the past, on the same three claims: any federal management program would thwart the transfer of public-domain lands to the states, the bill would close these lands to potential homesteading, and, for some livestock groups, the bill would be a catastrophe.

States' rights advocates had lost influence due to the failure of the Hoover proposal. Homesteading interests continued to oppose the bill, but by this time many members of Congress were coming to the conclusion that homesteading was already dying its own death due to the lack of remaining suitable land. There was great division within the ranks of the livestock industry. The favored position, almost universally, was that the lands should be ceded to the ranchers. If this option were closed, then the lands should be ceded to the states. This was also not a viable option at the time, and it was over the question of which alternative policy to pursue that livestock interests were split. It is unclear whether a majority of livestock ranchers opposed or favored the Taylor bill. It is clear that a number of organizations strongly opposed it, including the Arizona Cattle Growers Association, the California Cattlemen's Association, and the Wyoming Stock Growers.[7]

A number of factors contributed to the passage of the bill. It was a bill that had essentially already passed the House. It had been supported by the Hoover administration, and it was now supported by the Roosevelt administration. Representative Taylor was a strong and forceful advocate of the bill. And past supporters of such a policy—certain segments of the grazing industry and conservationists—added their support to the legislation. Among those livestock groups supporting the bill were the Colorado Wool Growers Association, the New Mexico Cattle Raisers Association, and the Utah Wool Growers Association. Livestock producer support could be traced to a variety of reasons, including a desire to see stability on the range, a thinking that the passage of the Taylor bill was inevitable, and a vision of the bill as an interim stage prior to state or private ownership (as Representative Taylor himself thought).[8]

Roosevelt's secretary of the interior, Harold Ickes, also publicly discussed taking executive action on the grazing lands if Congress failed to act. He refused to allow Civilian Conservation Corps workers or projects on public-domain lands until a grazing law was passed. Furthermore, he made a number of promises to stockmen to gain their support for the bill, and for Interior, rather than Agriculture Department, administration of the program. He promised not to create a large bureaucracy to implement the program and not to base fees on fair market value but on the cost of administering the program (for far less than the Forest Service). In testifying before Congress regarding the Taylor Grazing Act, Secretary Ickes stated, "We have no intention of making this a revenue producer at all. We would like for the range to pay for its own administration but nothing more."[9] There were also problems facing those who wanted to have the program administered by the Forest Service or another agency within the Department of Agriculture. Stockmen resented the Forest Service and its policies, especially in light of the continuing controversy over grazing fees in national forests. The stockmen preferred the creation of a new agency over which they could exert more control.[10]

The bill became law on June 28, 1934, despite the Department of Agriculture's recommendation that President Roosevelt veto it. The Taylor Grazing Act authorized the secretary of the interior to establish grazing districts on the public-domain lands and to develop any regulations necessary to administer such districts, including the granting of leases for up to ten years, the charging of fees (the bulk of which were to be returned to the districts for range improvement), the undertaking of range improvement projects, and the establishment of cooperative agreements with grazing landholders in the area. The secretary was also to cooperate with "local associations of stockmen" in the administration of the grazing districts, a feature of the act that Representative Taylor referred to as "democracy on the range" or "home rule on the range." Finally, the act included the phrase "pending final disposal." This was included because many of the supporters of the act thought it just that; the lands would eventually be transferred to the states or to private ownership. This phrase served to lessen the opposition of those who supported having the lands transferred to the private sector or the states.[11]

In September 1934, Farrington Carpenter, an attorney and rancher from northwestern Colorado, was appointed director of the new Division of Grazing. Carpenter very much supported the idea of "home rule on the range" and based his administration of the Taylor Grazing Act upon it. He sought to minimize the number of federal employees and to rely on local advisory boards in each grazing district to manage the program. These boards were to practice "home rule on the range," and in one letter he went so far as to write that "nothing but appellate power is expected to be reserved in the Government, thereby delegating to the committees practically the entire administration of the Act." One key responsibility of these boards was the assignment of grazing permits to ranchers. This was based on the ownership or control of commensurate private property or water rights, a process that essentially legitimized the status quo (and extralegal property rights where they existed).[12]

This structure established by Carpenter to administer the grazing districts had two significant consequences. First, it gave local grazing interests tremendous control over policy. According to Carpenter, "In practice their advice was followed in 98.3 percent of the cases." As Phillip Foss writes, "A small interest group has been able to establish a kind of private government with reference to federal grazing districts." National and state advisory boards concerned with more general policy issues supplemented the local boards. The National Advisory Board Council (NABC) president in the 1940s stated: "The revised [Federal Range] Code was written in its entirety by livestock men at the first meeting in Denver. The Grazing Service even asked if we would rather they weren't there." The NABC had very close relations with both the National Wool Growers Association and the American National Livestock Association, because most members of the NABC were members of one or both of the organizations.[13]

One telling example of this power took place in 1947, when the funds of the Bureau of Land Management (created by merging the General Land Office and the Grazing Service in 1946) for range management were drastically slashed by Congress. The advisory board contributed $200,000 of range-improvement funds for salaries in order to maintain some degree of administration on the lands. As Foss writes, "The payment of employees' salaries by advisory boards is probably unique in administrative history. In effect, the regulators were being supervised by those who were to be regulated." These advisory boards institutionalized interest-group liberalism into grazing policy and clearly demonstrate a captured policy regime. That government figures (e.g., Carpenter) and others supported this design of grazing policy, one that clearly furthered the interests of ranchers, indicates the acceptance of the idea of interest-group liberalism by some segments of society. Not only was the delegation of power to these local advisory boards acceptable; it was the preferred method to administer government policy.[14]

Second, the advisory boards reduced the potential friction involved in implementing the new program. If it were not for these boards, the program would have had no chance to be implemented. Indeed, the need for such boards was by design and is reflected in the low state capacity in grazing policy (especially in contrast to the Forest Service). The Division of Grazing was structured in such a way that it was destined to be a weak agency. The sources of weakness were both internal and external to the agency. Internally, battles within the Department of the Interior, low staffing and funding levels (a budget averaging $445,000 through 1940 and 205 employees in 1940 to administer 258 million acres and 12 million head of livestock), the lack of a profession of range management, frequent reorganization and decentralization, and the stockmen advisory boards all proved troublesome. The lack of state expertise and capacity led to grazing policy relying on societal expertise and capacity (ranchers), a recipe for capture.[15]

Externally, conflict with the Forest Service and Congress sapped much of the resources and energy of the Division of Grazing (later the Grazing Service) in the 1930s and 1940s. The jurisdictional battle contributed to the weakness of the Division of Grazing in three ways: it delayed the passage of grazing legislation; the Forest Service and its allies in Congress and the conservation community lobbied against placement of the program in Interior, attacked it once it was established, and attempted to have the program moved to the Agriculture Department; and the management policies of the Division of Grazing were often developed with this dispute in mind.[16]

The Division of Grazing/Grazing Service was in a difficult position in its relations with Congress as well as within the executive branch. Its only supporters in Congress were those representing the stockmen, and in order to continue to receive this support, the Grazing Service had to make sure its policies corresponded to the wishes of these stockmen. In the 1940s, congressional support lapsed as the Grazing Service was attacked by Senator Pat McCarran (D, Nev.) and his allies. Not only was the Grazing Service attacked by its only potential supporters; soon it was also under attack by the House Appropriations Committee as well. McCarran's forces attacked because grazing fees were too high; Appropriations attacked because the fees were too low. The Grazing Service, caught between these two congressional forces with no supporters to aid it, suffered drastic personnel and funding cuts in the late 1940s.[17]

In summary, passage of the Taylor Grazing Act resulted in policies that were controlled by the regulated interest group and the local elites that comprised the livestock industry, establishing a captured policy regime justified by the privileged idea of interest-group liberalism. Despite federal ownership, the grazing lands were managed as if they were private property: the low permit fees went primarily to range improvements, and the grazing rights attached to a particular piece of private property became capitalized within the value of that property.

The Federal Land Policy and Management Act: An Organic Act for the Bureau of Land Management

The BLM had been operating on insecure foundations since its creation in 1946. Unlike its predecessor agencies, the Grazing Service and the GLO, the BLM did not have a congressional mandate. Rather, it had been created by an executive reorganization; hence, it could be reorganized out of existence by another executive reorganization. More important, the BLM did not have any legislative charter or mission. It was a caretaker agency practicing custodial management on lands the future of which was uncertain.

The movement to change the status of the BLM and its lands was initiated by the agency itself. In 1960, the Multiple Use and Sustained Yield Act gave the Forest Service statutory authority to continue the type of management philosophy it had been employing on national forests. The BLM, with its collection of land managers and growing collection of resource professionals committed to technocratic utilitarianism, sought to have Congress enact similar legislation to guide the agency. Bureau officials were also encouraged by the election of President Kennedy, who was concerned with natural resources issues. In response, agency officials drafted a bill in 1961 to modernize public-land laws, its first effort in this direction. This effort continued for fifteen years, as the BLM was the major source of pressure to institute an organic law.[18]

Two laws passed in 1964 signaled the beginning of significant movement toward an organic act. The Public Land Law Review Commission (PLLRC) Act established a commission to review the vast number of laws dealing with public lands and make recommendations to improve public-lands management. This commission was the idea of House Committee on Interior and Insular Affairs chair Wayne Aspinall (D, Colo.), who traded his support for the wilderness bill for the creation of the committee (see Chapter 4). Also in 1964, the Classification and Multiple Use Act directed the BLM to classify the lands it managed for retention or disposal. This was the first law authorizing the BLM to inventory its lands and resources, suggesting that some of these lands would continue in federal ownership.[19]

The PLLRC delivered its report, *One Third of the Nation's Land*, in 1970, after six years of hearings and information gathering. Aspinall chaired the commission and the report reflected his position. The report favored a dominant-use, in contrast to a multiple-use, approach and increased use of the public lands for commodity production, that is, timber, forage, minerals, energy. Little change was recommended in current grazing practices on public lands. Overall, Milton Pearl, commission director, stated that "the commission emphasis is on letting the market system determine when resources should be sold." Aspinall claimed that the report represented the true conservation philosophy: "the maximum good for the maximum number."[20]

These comments reflect the dominance of economic liberalism in the PLLRC report. However, it was primarily those supporting the status quo—interest-group liberalism—that spoke for commodity interests. Ranching interests operated under a satisfactory system with federal ownership: they were charged low fees, had relatively secure tenure to the lands that they leased, and had a great deal of input into the determination of grazing policy. In the discussions of an organic act for the BLM, livestock interests took the position of favoring the status quo, and if possible, extending the use of the public lands for commodity production.

Although the report received favorable reviews from commodity users, it was attacked by environmentalists. Sierra Club representatives claimed that "the basic assumption of the Report is that we will continue to have the no deposit—no return, we-use-once-and-throw-away, Philistine culture we have today" and that "the Report is not balanced at all. In fact, the pattern is one of preserving maximum advantage for western commercial interests." The National Wildlife Federation commented that the report was "carefully couched to give the illusion of maximizing the public benefits from Federal lands, although its primary thrust is to give commercial development and use preeminence over recreation, esthetics and related values. In 1930, such recommendations would have been unacceptable to the American public. In 1970, they are incredible."[21]

Almost as soon as the report was made public, Congress began to work on bills to implement the recommendations of the PLLRC. In 1971 and 1972, three general types of bills were under consideration. The first type, introduced by Representative Aspinall, provided a general framework for future, more specific legislation. This approach was generally favored by commercial users, including the American Farm Bureau Federation, American Mining Congress, and National Forest Products Association.[22] A second type of bill, sponsored by Senate Interior and Insular Affairs Committee chair Henry Jackson (D, Wash.), provided the BLM with an organic act and included sections reforming the 1872 Mining Law and providing environmental safeguards for BLM activities. This bill was supported by environmental and conservation interests, chiefly the Public Lands Conservation Coalition, consisting of sixteen environmental and conservation groups. Preservationists supported the BLM in its efforts to achieve an organic act due to their desire to see the public-domain lands retained and to see the agency manage these lands for multiple use. This coalition with the technocratic utilitarians is clear in the Public Lands Conservation Coalition, which included preservationists and those supporting the technocratic utilitarianism idea.[23] The third type of bill under consideration, one drafted by the administration and the BLM, provided for a BLM organic act without attempting major reforms of other public-land laws. The technocratic utilitarians wanted a bill only to achieve clear ends: BLM statutory management authority and the retention of public domain lands. It opposed the addition of other sections to any bill.[24]

In the Senate, the Jackson bill made it out of committee, and in the House the Aspinall bill made it out of committee, but both bills died with the Ninety-second Congress before they could be considered. Additionally, bills were introduced in both chambers to amend the Taylor Grazing Act to allow for a statutory fee formula based on a rancher's ability to pay and the comparative value of forage. Neither bill passed.

The Ninety-third Congress opened without the major actor in public-lands policy. Representative Aspinall was defeated in the 1972 Democratic primary, due largely to the work of environmental groups. In his absence, there was a lack of strong leadership on public-lands issues in the House. Representative John Melcher (D, Mont.), new chair of the Subcommittee on Public Lands, took over work on the PLLRC recommendations. Melcher and his colleagues drafted a new bill, based on extensive public hearings, which contained a section on grazing that addressed such controversial topics as grazing fees and rancher tenure. Ranchers decided that they, too, should gain something from this act, and these provisions would improve their already favorable status quo. The addition of grazing issues to the organic bill suggested future difficulties for the BLM, because the bill would no longer be a simple, relatively noncontroversial organic act.[25]

An organic bill was faring better in the Senate at this time. Despite some differences regarding the classification of wilderness on BLM lands, a bill introduced by Jackson passed the Senate overwhelmingly, seventy-one to one. This bill was similar to Jackson's bill in the Ninety-second Congress, with the exception that the controversial mining section was modified. For a time, it appeared that an equivalent of the Jackson bill would pass in the House, a bill sponsored by Morris Udall (D, Ariz.) and backed by the Nixon administration. After being lobbied by the United States Chamber of Congress, the Liberty Lobby, and various natural resources corporations, Nixon, however, in need of conservative support during the Watergate crisis, withdrew his support. The Udall bill was denied a rule by the House Rules Committee and did not reach the House floor. In the meantime, the Melcher subcommittee bill was still being worked on in committee when the Ninety-third Congress adjourned.[26]

In 1975, Jackson reintroduced a bill nearly identical to the one that had passed the Senate earlier. Senate consideration of the bill was delayed to give the House time to work on its own bill, but with little progress in the House, the Senate easily passed Jackson's bill again in February 1976. There was an effort on the floor, however, to include an amendment to establish a statutory grazing fee formula. The amendment, strongly opposed by Jackson, the administration, and environmental groups, was defeated. The Senate was growing weary of waiting on the House, however, and one senator went so far as to claim that this was the last chance for an organic bill: "The committee would not likely take up this bill for a fourth time in the next Congress."[27]

The House began the Ninety-fourth Congress considering three main bills, much as it had in the Ninety-second Congress. There was an administration bill, a bill sponsored by Representative John Seiberling (D, Ohio) that was favored by many environmental groups, and a subcommittee draft bill. The subcommittee put most of its energy into revising the subcommittee bill, drafted by subcommittee members Melcher, Sam Steiger (R, Ariz.), Jim Santini (D, Nev.), and James Weaver (D, Ore.). At hearings held in the spring of 1975, the subcommittee heard testimony on all aspects of the bill, but especially on the bill's grazing provisions, which were not part of the Senate bill. The chief grazing issues included in the bill were: (1) the establishment of a statutory grazing fee formula based on beef prices and private forage, (2) the improvement of rancher permit tenure by granting permits for ten years as a rule rather than as an exception, (3) the compensation of ranchers for private improvements of the land if permits were canceled, and (4) the existence and composition of grazing advisory boards. Commodity users supported the bill, and environmental groups and the administration opposed it.[28]

Grazing interests, represented by the American National Cattlemen's Association, the American National Wool Grower's Association, and the Public Lands Council, favored almost all aspects of the bill. The only grazing aspects of the bill that they opposed were the specific fee formula (because it was tied to private forage costs), the $2.00 per AUM grazing fee floor (which they termed "arbitrary"), and the grazing allotment plans (because bureaucrats would be telling ranchers how to ranch). Both the American Mining Congress and the National Association of Counties favored the bill in general, although both expressed reservations concerning wilderness issues. The American Farm Bureau Federation gave the bill tacit support but only after advocating the transfer of these lands to private ownership.[29]

Opposition to the subcommittee draft was of three varieties. First, the administration opposed the bill, supporting its own bill instead. Bureau of Land Management director Curt Berklund testified in favor of a bill with a broad policy statement rather than specific directions on resource use. He was opposed to the changes in grazing law included in the bill, which he argued were misplaced. This was to be a BLM organic act; if Congress wished to amend grazing laws, it should amend the Taylor Grazing Act. The chief of the Forest Service, John McGuire, also testified in favor of the administration's organic bill. He expressed the Forest Service's opposition to the grazing sections, because the subcommittee draft bill specified that these provisions would apply to the Forest Service as well as the BLM. McGuire was especially critical of the section on compensation for personal improvements on public lands if permits were canceled: "But we are even more concerned about the implication that this privilege of the permittees could be interpreted as a property right. You know, there has been a long debate about the definition of a grazing permit, whether it is a privilege or a property right, and I believe that is our basic fear, if this compensation requirement became law."[30]

A second variety of opposition came from environmental groups that opposed the subcommittee draft bill and favored the Seiberling bill. A panel of eleven conservation groups testified in support of a BLM organic act, with the recommendation that the Seiberling bill be used for the markup. They opposed all of the sections dealing with grazing in the subcommittee draft bill. At times, the Sierra Club also was part of this coalition, but in its final testimony the Sierra Club argued for the creation of a new land-management agency: "The lands need an organic act—not BLM. Consequently, we urge the Committee to consider creating through this legislation an entirely new agency."[31]

The third type of opposition to the subcommittee draft bill came from preservationist groups opposed to all three bills. These groups focused on the BLM's new role in energy management, which the preservationists thought ran counter to any hopes for good, sound environmental management by the BLM. The Wilderness Society opposed the grazing provisions in the subcommittee draft bill, especially regarding tenure and compensation: "[The bill] comes close to establishing a vested right, and it gives a specially privileged status to grazing permittees that is enjoyed by no other users of the public lands. . . . We recommend against any language that would allow compensation for the value of cancelled grazing permits, or any similar compensation." The Wilderness Society representative continued: "As presently organized, the Bureau of Land Management is fundamentally incapable of managing the public lands under conservation principles. . . . [There is a] fundamental conflict between the mandate of this organic act and the mandates of the energy exploitation laws. . . . We believe a basic reorganization is needed." Friends of the Earth took a similar position in its testimony, favoring one division of the BLM for surface management and one for subsurface management, and opposing all of the bills being considered.[32]

During and directly after these hearings in the spring and summer of 1975, it appeared that the House subcommittee was making good progress on its bill. Even the controversial grazing-fee section was determined relatively smoothly. A partisan breakdown of the subcommittee soon followed, however, and work on the bill slowed considerably. By the spring of 1976, the subcommittee's bill was nearly completed, and it was strongly opposed by environmental interests. The Sierra Club argued: "It is plain from this and other parts of the bill that the House subcommittee considers the BLM Organic Act not as a basic charter for new and more progressive management of the public domain lands, but rather as a vehicle on which to hang special favors to the grazing and mining industry. . . . The outlook for the BLM Organic Act is grim." The preservationist-technocratic utilitarian coalition dissolved. Preservationists became more interested in the addition of preservation features to the bills, such as the review of BLM lands for wilderness and the preservation of significant portions of the California desert. A bill that did not include sufficient language dealing with these issues would be unacceptable to the preservationists. In addition, they became more critical of BLM activities.[33]

Despite this opposition, the subcommittee reported the bill to the full Committee on Interior and Insular Affairs. A few minor changes were made there, and the bill was reported out of committee by a narrow 20 to 16 vote in May 1976. This vote foreshadowed a tight floor vote, and, indeed, the bill passed 169 to 155 in July 1976. Those opposing the bill were generally influenced by the opposition of environmental groups and the Department of the Interior's skepticism. Amendments were introduced to delete the section on grazing fees and to force the BLM to charge the fair market value for these fees, but both amendments were soundly rejected.

Over twelve other proenvironmental amendments were adopted on the floor, however, making the bill more palatable to environmentalists.[34]

Although both the House and the Senate had now passed versions of a BLM organic bill, there were major differences in the bills and not much time left before adjournment. A number of significant issues needed to be worked out, including law enforcement, grazing fees, grazing tenure, grazing advisory boards, wild horses and burros, unintentional trespass, and the California Desert and King Range Conservation Areas. By September 22, four key differences between House and Senate conferees remained: (1) the grazing-fee formula, (2) the standard ten-year grazing permits, (3) the grazing advisory boards, and (4) a section making changes in the patenting of mining claims. The first three items were contained in the House bill and opposed by the Senate conferees, the last item was in the Senate bill and opposed by the House conferees. Senator Lee Metcalf (D, Mont.) offered a compromise in which all four controversial sections would be dropped from the bill. The Senate conferees approved the compromise, but the House conferees split five to five. The deadlock was primarily over the statutory grazing fee formula: the Senate opposed any such fee, the House conferees (at least five) held that a fee formula must remain in the bill. With Congress set to adjourn on October 1, it appeared that the bill would die in conference.[35]

Over a weekend break, last-ditch efforts were made to salvage the bill. The mining interests knew that the Senate would drop the patenting language it objected to if the grazing-fee provision were dropped, so the American Mining Congress pressured the livestock interests to allow the formula to be dropped, but they balked at this. At the next meeting, the House offered a compromise: to use its fee formula for two years, then to deal with the issue again. The Senate refused. The House then offered to drop the formula and substitute a two-year freeze on grazing fees. This, too, was rejected by the Senate. The Senate conferees made a final proposal: a one-year freeze on grazing fees and a one-year study of the fee issue by the Departments of Agriculture and the Interior. Once the study was completed, Congress could then deal with the issue. In addition, the Senate would accept the House position on the other three troublesome issues: permit tenure, advisory boards, and no change in mining patent limits. Both sides agreed to the compromise, and the bill was voted out of the conference committee. There were rumors that disgruntled livestock interests would attempt to have the bill killed on the floor of the House, but despite the rumors, once the Interior Department informed Congress that it favored the bill, it comfortably passed the House on September 30 and the Senate on October 1, just hours before the Ninety-fourth Congress adjourned.[36].

The battle was not quite over, however. Grazing interests were displeased that the grazing-fee formula had been dropped from the bill, and they urged President Ford to let the bill die via a pocket veto. President Ford followed the advice of his Interior Department, though, and signed the bill on October 21, 1976.[37] As enacted, the Federal Land Policy and Management Act (FLPMA) of 1976 included important policy provisions regarding the BLM, grazing, and preservation. The act declares that "the public lands be retained in public ownership," which made it clear that public-domain lands would be retained. This had been the policy since the passage of the Taylor Grazing Act, but the "pending final disposal" phrase in that act made federal tenure and management decisions concerning these lands tenuous. Title III of the act was essentially the BLM organic act. It stated that the BLM would perform the duties and functions vested in the secretary of the interior relating to the administration of certain laws. It also directed the president to appoint a director of the BLM, to be confirmed by the Senate. Part of the BLM mission would include the practice of long-range, multiple-use planning, similar to that done by the Forest Service.

There are three important sections of the FLPMA concerning grazing management for both the BLM and the Forest Service. As discussed above, the conference committee compromise dictated a freeze on grazing fees in 1977 and a one-year study of the issue undertaken by the Departments of Agriculture and the Interior. A second section directs the BLM and the Forest Service to issue ten-year permits for grazing, with specific exceptions permissible. If at the end of that ten-year period the lands are to remain available for grazing, the current permittee has the first priority for lease renewal (pending his or her having met the conditions of the lease). Thus, this section virtually guarantees a rancher the use of certain public lands as long as those lands are to be used for grazing. It is also stated in this section that the permittee will receive reasonable compensation for private improvements on the public land if his or her permit is canceled at any time. The importance of property rights vested in grazing permits is apparent in these two sections. The third important grazing section concerns grazing advisory boards. These boards, established after the passage of the Taylor Grazing Act and given a statutory base by a 1939 amendment to the act, were eliminated by law in the early 1970s, replaced by multiple-use advisory boards. The FLPMA gave new statutory life to grazing boards (but did not eliminate the multiple-use boards). The purpose of the boards was to advise the BLM and the Forest Service on allotment management plans and range-improvement funds. Like the earlier boards, members were to be selected by election from permittees in the particular area.

Finally, the FLPMA included several significant preservation policies. It established a California Desert Conservation Area, which recently became home to three new national parks and nearly eight million acres of wilderness. The act also includes provisions for the review of BLM lands for wilderness designation, because these lands were not included in the original Wilderness Act (due to the focus on what were perceived to be the more important national forest, national park, and wildlife refuge lands). The controversial BLM wilderness review and designation process is now well under way.[38]

Clearly, the BLM was pleased to finally have an organic act. The FLPMA gave the bureau a statutory foundation and a mandate to manage its lands, lands that would be retained. The implementation of the FLPMA has been problematic, though, due chiefly to the lack of capacity of the BLM. The agency has been hampered by poor leadership and has been more susceptible to dispositions of political appointees. A second problem has been the lack of a consistent, qualified, professional workforce. There are a great many entrenched employees who are much more inclined toward livestock and mineral production than conservation or multiple use. George Coggins argues that "BLM personnel tend to have narrow educations, narrower political viewpoints, and an encompassing ignorance of the law which should govern their operations. . . . Overstrained range conservationists all too often use ad hoc, biased, and unscientific approaches in the field." Due to the FLPMA, the BLM also underwent growing pains when it had to hire a variety of new natural and social scientists to implement the provisions of the act. This inflow of new and different personnel has not been easy to absorb. It has, however, increased the professional stature of the agency and its capacity and has helped bolster the idea of technocratic utilitarianism within the agency.[39]

Another problem contributing to the weakness of the BLM as an agency has been its poor standing within the Department of the Interior. According to former BLM director Frank Gregg, the BLM has been "historically weak in sustaining its decisions within the Interior Department." There also appears to be a lack of support within the department for a strong and vibrant BLM. The agency has also had poor standing within the past few presidential administrations. Following the passage of the FLPMA, with the BLM poised to begin a life of professional land management, Interior Secretary Cecil Andrus described the agency as the Bureau of Livestock and

Table 4. Administrative Capacities of the Bureau of Land Management
and the Forest Service, by Fiscal Year

	BLM	FS
Employees		
1978	7,000	44,000
1983	10,148	30,499
1988	12,725	32,443
Budget ($)		
1978	440 million	1,190 million
1983	507 million	1,621 million
1988	641 million	1,628 million
Acres managed		
1978	480 million	189 million
1983	342 million	192 million
1988	334 million	191 million

Source: Office of Management and Budget, *Budget of the United States, Fiscal Year 1988 (and Appendix)*; U.S. Department of the Interior, Bureau of Land Management, *Public Land Statistics 1987, Public Land Statistics 1984, Public Land Statistics 1978*.

Mining. The Reagan administration sought to undermine BLM management through budget cuts and changing regulations. A fifth problem leading to BLM weakness has already been discussed: the pattern of capture that dates back to the Taylor Grazing Act and the 1930s and 1940s. This history "continues to haunt the agency." Ironically, the BLM has also "been unable to enlist its rancher clientele as supporters."[40]

Without question though, the most important factors contributing to the weakness of the BLM is its lack of funding and lack of personnel, weaknesses that are especially apparent in comparisons with the Forest Service (see table 4). These factors make it extremely difficult for the BLM to manage its lands, making the agency more reliant upon ranchers, because their cooperation is required to implement these programs.[41]

To summarize, most analysts blame the lack of capacity of the BLM for the failures in the implementation of the FLPMA. Coggins writes that "probably the single major problem in public rangeland management is the managing agency. . . . The BLM is a travesty of a land management agency." "It should come as no surprise, then," writes John Baker, "that the Bureau's institutional weaknesses show up in the conduct of programs mandated by its 'organic act.' "[42]

Another aspect of state structure that contributes to the difficulties of the BLM and grazing management is the congressional committee system. Throughout the discussion leading to the passage of the FLPMA, the interests of commodity users, especially grazing interests, received strong representation on the House Public Lands Subcommittee. This subcommittee was dominated by westerners (illustrative of the captured policy pattern), and the policy positions these members expressed were not representative of the entire Congress. Because the subcommittee controlled the legislation, however, the bill that emerged was more pro-use and less environmentally oriented than Congress as a whole would have favored. It was either accept these compromises or have no organic bill at all. This bias also helps to explain the disproportionate political power of the less than thirty thousand ranchers holding grazing permits for BLM lands. Although a very small number nationally, these ranchers are key political actors in many sparsely populated western states.

One consequence of the act and the planning and management that it has required was the stimulation of western hostility to the BLM and the existence of so much federal lands. According to a number of analysts, the FLPMA was "the match that finally lit the Sagebrush Rebellion," the movement to have federal lands transferred to the states. Another problem has been the legislative veto provisions built into the act. Because legislative vetoes have been found unconstitutional, the constitutionality of certain major portions of the act are in question.[43]

The FLPMA is a crucial piece of legislation in terms of the BLM and in terms of public-lands grazing management. It gave the BLM statutory authority and a mandate for multiple-use management. The act itself was a victory for technocratic utilitarianism and included preservationist provisions. Although it did not displace the privilege of interest-group liberalism in the grazing policy regime, the FLPMA allowed the other two ideas to gain a footing. Nevertheless, although the supporters of interest-group liberalism were unable to block passage of the bill and maintain an unchanged status quo, the gains of the technocratic utilitarians and preservationists were less substantial than they might seem. Livestock interests were able to reduce professional discretion on grazing due to the provisions included in the act, and the FLPMA did not designate any wilderness; it only began a review process. Implied property rights for ranchers were further strengthened in the act. And the BLM remained underfunded and understaffed. Hence, though some blows to interest-group liberalism had been struck, its privilege and the captured policy pattern remained.

The Dispute over Grazing Fees in the 1970s and 1980s: Economics, Politics, and Wildlife

From the beginning of government management of grazing on public lands, both on the national forests and the public-domain lands, the fees charged for grazing permits has been a controversial issue. When the Forest Service began charging such fees in 1906, the policy was vehemently opposed by cattlemen and sheepmen, who challenged its constitutionality. Later, Forest Service efforts to increase fees in the 1920s contributed to rancher opposition of Forest Service administration of public-domain grazing.

With the passage of the Taylor Grazing Act, fees for grazing on public lands (nonnational forest lands) were initiated at $.05 per cow or horse per month (or per AUM) and $.01 per sheep or goat per month. This fee level appeared reasonable due to the former free use of the range, the Depression, poor range conditions due to drought, the cost-of-administration basis for the fees promised in the debate over the Taylor Grazing Act, and because the Forest Service was charging $.08/AUM (see table 5 for the fees charged by the Grazing Division/Grazing Service/BLM and the Forest Service from 1935 through 1985). This fee did not rise, however, until 1947. In the early 1940s, the House Appropriations Committee began to demand higher grazing fees. The Grazing Service, though, was receiving pressure from livestock groups to keep the fee at $.05/AUM, pressure applied especially through Senator McCarran and the Senate Public Lands Committee. In 1944, the director of the Grazing Service recommended tripling the fee to $.15/AUM, which was strongly opposed by the National Advisory Board Council and McCarran. In 1946, McCarran's committee reported that there was no justification for any fee increase. Clearly, the Grazing Service was not in a good political position. The budget for the Grazing Service was halved, and the service was merged with the GLO to form the BLM.[44]

In 1946, the Nicholson Report, commissioned by the secretary of the interior, was issued. The report recommended raising the fee to $.08/AUM based on a cost-of-administration approach. This fee was implemented in 1947, despite opposition by some members of Congress and Interior Department officials who favored a higher fee. In 1950, the fee was raised to $.12/AUM to cover inflation and the addition of more BLM employees. The NABC approved a change, in 1954, in the basis of the fees from the cost-of-administration approach to a method based on the price of livestock. This

Table 5. Bureau of Land Management and Forest Service Grazing Fees, 1935–1985

Year	BLM ($ per AUM)	FS ($ per AUM)	Year	BLM ($ per AUM)	FS ($ per AUM)
1935	—	0.08	1961	0.19	0.46
1936	0.05	0.13	1962	0.19	0.46
1937	0.05	0.13	1963	0.30	0.49
1938	0.05	0.15	1964	0.30	0.46
1939	0.05	0.13	1965	0.30	0.46
1940	0.05	0.15	1966	0.33	0.51
1941	0.05	0.16	1967	0.33	0.56
1942	0.05	0.19	1968	0.33	0.56
1943	0.05	0.23	1969	0.44	0.60
1944	0.05	0.26	1970	0.44	0.60
1945	0.05	0.25	1971	0.64	0.78
1946	0.05	0.27	1972	0.66	0.80
1947	0.08	0.31	1973	0.78	0.91
1948	0.08	0.40	1974	1.00	1.11
1949	0.08	0.49	1975	1.00	1.11
1950	0.08	0.42	1976	1.51	1.60
1951	0.12	0.51	1977	1.51	1.60
1952	0.12	0.64	1978	1.51	1.60
1953	0.12	0.54	1979	1.89	1.93
1954	0.12	0.35	1980	2.36	2.41
1955	0.15	0.37	1981	2.31	2.31
1956	0.15	0.35	1982	1.86	1.86
1957	0.15	0.34	1983	1.40	1.40
1958	0.19	0.39	1984	1.37	1.37
1959	0.22	0.50	1985	1.37	1.37
1960	0.22	0.51			

Source: Secretary of the Interior and Secretary of Agriculture, *Study of Fees for Grazing Livestock on Federal Lands*, pp. 2-4, 2-5, 2-29; Williamson, "Where the Grass Is Greenest," pp. 30–31.

system was gradually phased in, with fees reaching $.19/AUM in 1958. It should be noted that by this time, the gap between the fees charged on BLM and Forest Service lands had grown considerably, with Forest Service fees double BLM fees in 1958 (see table 5).[45]

The comptroller general and Bureau of the Budget entered the scene in 1959, recommending that both BLM and Forest Service grazing fees be raised so that they would approximate a fair market value (FMV) for the forage. The issue took on increased importance when it was mentioned by President Kennedy in his message to Congress on natural resources in 1962. In response to these executive-branch policy positions, the Interdepartmental Grazing Fee Committee, consisting of representatives from the Departments of Agriculture, Defense, the Interior, and the Bureau of the Budget, was formed in 1960. The report of the committee, released in 1967, included three major conclusions: (1) that a uniform base should be used by all federal agencies in establishing fees, (2) that fees should be based on the economic value of the use of the public lands to the users, and (3) that the economic value should be such that the government receives a fair return.[46]

At this time, the grazing fee debate focused on the ideas of economic liberalism and interest-group liberalism. The latter idea accepted government ownership of the public lands because the system benefited the ranchers who used the public lands in their operations. They had developed a productive working relationship with the federal government, and ranchers favored a continuation of this status quo. Those adhering to economic liberalism, although not always supporting the transfer of public grazing lands to the private sector, consistently favored raising fees to the FMV, that is, having fees based upon information gleaned from free market activities rather than established based on some bureaucratic standard.

Another government study, the Western Livestock Grazing Survey, was begun in 1966. This survey gathered information on ranching costs (such as lease rates, nonfee costs, and market value of grazing permits) from approximately ten thousand people. This information was used by the special interagency Grazing Fee Technical Committee, which issued a report in 1968 (the Houseman Report) that stressed: (1) that there were no statistical supports for different base fees between BLM and Forest Service lands; (2) that the statistical differences were not significant enough to justify different fees in different areas; and (3) that the FMV was based on the lease rate for private forage adjusted by the difference in nonfee costs on private and public lands (nonfee costs were higher on public lands). The fees being charged at the time were $.33/AUM by the BLM and $.56/AUM by the Forest Service.[47]

The Bureau of the Budget announced that both BLM and Forest Service fees would be increased to achieve the FMV figure of $1.23/AUM. To mitigate the hardship on ranchers, the $.90/AUM increase for BLM permittees and $.72/AUM increase for Forest Service permittees would be phased in over ten years. The new formula was also indexed to the changing rates on private lands.[48]

Hearings were held before the public-lands subcommittees in both the House and Senate in 1969. Predictably, livestock interests opposed the new fee formula. They argued that the permit value should be included as a cost factor, because they had to pay more for private land with public-land grazing privileges. They also cited the adverse effects of such increases. Conservation groups favored the fee increase, since it would lessen the subsidy to the ranchers and increase the funds available for range-improvement programs. The American Forestry Association, for example, applauded the increase in an editorial: "AFA warmly lauds Mr. Udall and Mr. Freeman for what amounts to a new Declaration of Independence on the part of agencies grazing on public land. We are all for it."[49]

The new fees were quickly challenged in court. In a case in New Mexico, brought by the chair of the forest committee of the American National Cattlemen's Association, the plaintiff argued that grazing permits were protected under the Fifth Amendment to the Constitution, and the increase in fees for the permits violated this protection. This case was dismissed by the district court, which upheld the lower court decision: the vested interest in grazing permits does not legally exist. Another case in Utah also upheld the fee increase.[50]

To many ranchers, the grazing permits (and the low fees attached to them) were a property right. Ranchers were willing to pay more for private lands that had associated public grazing privileges, and soon, the value of these permits became capitalized into the value of the private lands. These permit values became a type of property right in the eyes of many in the West, including ranchers, bankers, and economists. They were used as security for bank loans and included in the appraised value of ranches. True, a gift equal to the value of the first permit was granted to the first private landholders who received permits. But future landholders received no gift; they paid for the permit by paying more for the private lands they purchased. To raise the fees to eliminate the permit value would mean double-charging current permit holders. One economist argued that any large increase in fees would not be equitable "since it would reduce the private capitalized value of the ranch, and would, in fact, be taking private property without compensation."[51] Some economic liberals went further, arguing that the problems with grazing on public lands were due to a lack of formal and secure property rights for ranchers. Debates over the proper grazing fees were a waste of effort. The only way the problems associated with low grazing fees could be solved was by making the property rights formal and secure through transfer to the private sector.[52]

A second increase in fees was scheduled for 1970, but it was postponed. Secretary of the Interior Walter Hickel declared a moratorium on a fee increase in 1970, claiming that such an increase should be postponed until the PLLRC issued its report. Hickel was not acting alone on this, however: a group of nine western senators pressured the secretary for the moratorium, claiming that the fee increase would put marginal ranchers out of business.[53]

Over the following seven years, four full-scale increases in fees were made. In 1971, BLM fees were raised $.20/AUM, $.09 for FMV reasons and $.11 to cover the increased value of forage on private lands. Increases were limited to only $.02/AUM in 1972 by Secretary of the Interior Rogers Morton in order to keep fee increases to 3 percent in support of the Economic Stabilization Program. Full fee increases were made in 1973 and 1974, with BLM fees increasing $.12/AUM and $.22/AUM (again, to reflect the phase-in to FMV and the increase in the value of private forage).[54]

Another moratorium was declared in 1975 by Interior Secretary Morton and Agriculture Secretary Earl Butz. They cited the "difficult economic and drought conditions facing much of the livestock industry throughout the Western states." It should be noted that the drought of 1975 was not a significant one, and that no aid was made available to ranchers on private lands. Rates were again increased in 1976, by $.51/AUM on BLM lands. In order to achieve the goal of FMV by 1980, the annual nonadjusted BLM increase was boosted from $.09/AUM to $.11/AUM. Yet another moratorium was put in place for 1977, as part of the FLPMA.[55]

Recall that during the early 1970s much debate on grazing policy and grazing fees in particular focused upon the BLM organic bills. The livestock interests attempted to have a statutory grazing-fee formula, based on the price of beef and private land forage, placed in the act. The House bill included such a provision, but it was dropped in the conference committee due to Senate objections. The FLPMA mandated a one-year moratorium on fee increases and a one-year study of grazing fees.

In the fall of 1977, Interior Secretary Cecil Andrus and Agriculture Secretary Bob Bergland proposed raising both BLM and Forest Service fees to a uniform $1.89/AUM for the 1978 grazing season. The fees would then rise at no more than 12 percent per year until reaching a FMV level of $2.38/AUM. The proposed increase was attacked by livestock groups. In a joint statement released by the National Cattlemen's Association and the National Wool Grower's Association, the proposed increases were called "unfair and unrealistic." Administration sources countered that based on the current formula, the fees would rise to $2.09/AUM for BLM lands and $2.15/AUM for Forest Service lands the next year, so ranchers were receiving a break.[56]

Also in the autumn of 1977, the joint Interior-Agriculture grazing-fee study, which analyzed seven different fee systems, was completed. After examining the various formulas, the report ranked the alternatives based on the analysis: (1) the modified current system, (2) the current system (adopted in 1969), (3) the 1976 technical committee proposal, (4) competitive bidding (ranked so low due to the inaccuracy of ten-year bids), and (5) National Cattlemen's Association, House Interior Committee, and American Farm Bureau Federation (AFBF) proposals. These findings were reflected in the administration proposal made public in October 1977.[57] Under pressure from ranchers, with the new grazing report just released, and with Congress likely to consider grazing issues in the next Congress, the Carter administration decided to withhold increasing fees until Congress decided on proposed new fee rates. Until Congress acted, the 1977 fees would remain in place. Just to prevent the administration from changing its mind, livestock interests pressed Congress to pass a one-year moratorium on fee increases in June 1978.[58]

With the fee moratorium in place, Congress focused its energy on grazing issues for the second time in three years. The bill that emerged from Congress in the fall of 1978 focused on range improvements and a new statutory grazing-fee formula. The sections of the bill dealing with range improvement had almost universal support from ranchers, from environmentalists, and from the BLM and the Forest Service. No one lost with more money for range improvement. There was great controversy over the grazing-fee formula, however. The BLM, Forest Service, and environmental groups favored the modified current system, based on the analysis contained in the report released the year before. The livestock industry supported the technical committee proposal, with its built-in guaranteed profit. After bitter debate in both the House and the Senate, the technical committee proposal was adopted, and signed by President Carter on October 25, 1978.[59]

The supporters of technocratic utilitarianism (namely the employees of the BLM and the Forest Service and range management academics) were institutionally crucial actors in the grazing fee debate but did not view the issue to be as important as other actors did. The greatest concern with fees was that increased fees usually meant increased funds for range management, so the technocratic utilitarians tended to favor fee increases. Unlike the preservationists, however, they did not think that grazing fees and the condition of the range were linked in a more direct way. Perhaps the BLM and the other main group of technocratic utilitarians, range management academics, never made such a strong stand for higher grazing fees because such an increase in fees would alienate their major, and in the past at least, only, clientele, the ranchers.

Preservationists had become involved on grazing fees because of their concern over how the fees were affecting animals, plants, and land that they were trying to preserve. They were concerned on two counts: (1) the low fees subsidized commodity-interest use of the public lands, and (2) they encouraged overgrazing that hurt the land, hurt fish habitats, and reduced the forage available for wildlife. Edward Abbey went so far as to refer to public-lands ranchers as "nothing more than welfare parasites" and advocated an end to all grazing on the public lands.[60]

The Public Rangelands Improvement Act (PRIA) declared the technical committee fee formula equal to FMV, with two conditions. First, the annual increase or decrease in the fee could not exceed 25 percent of the fee in the previous year. And second, the formula would only be in place for seven years, from 1979 through 1985. After 1985, Congress would have to wrestle with the issue again. The new formula went into effect in March 1979, and the fees were set at $1.89/AUM for BLM lands and $1.93/AUM for Forest Service lands. The increases, the first since 1976, were limited by the 25 percent maximum increase. Without this cap, the fees would have been raised to $2.03/AUM. The fees rose again in 1980, then fell in 1981—but for the first time BLM and Forest Service fees were the same. From this time on, both agencies would have the same fees. Each year through 1985 the fees continued to fall, reflecting poor beef prices, reaching $1.37/AUM in 1985 (see table 5).[61]

With the passage of the PRIA, the grazing fee issue was placed on the back burner as all concerned parties readied themselves to enter battle over the issue once again in 1985. In the interim, the Reagan administration followed policies favorable to livestock interests. The BLM began a cooperative management program (CMP) in which the government delegated almost complete management control of grazing lands to particular ranchers (reminiscent of "home rule on the range"). The federal government discontinued its push for a blanket claim to all water on federal lands; the government even encouraged ranchers to file for water claims on their leased federal land under state water law. Lastly, the BLM drew back from its allotment management strategy to significantly decrease the number of cattle grazed on public lands. At the same time, the privatization initiative was under way within the administration. Proposals including the disposal of millions of acres of grazing lands and doubling or tripling the fees charged on the remaining lands were made and discussed.[62]

Environmentalists were not slow to react to what they viewed as abuses of grazing policy by the Reagan administration. Former BLM director Frank Gregg attacked the CMPS: "What we are talking about here is a policy that amounts to a give-away of the public lands to private interests." The Natural Resources Defense Council (NRDC) filed suit against the CMPS in 1984 and was successful in having the policy halted. Environmentalists also attacked the policy of subletting permits on BLM lands. A rancher would receive his or her permits from the BLM, and then sublet them to other ranchers, often for fees as high as ten times the government fee. It is estimated that as many as 50 percent of BLM permit holders engaged in subletting, a practice that is illegal on Forest Service lands. In response to this environmentalist pressure, the Interior Department prohibited the subleasing in 1985.[63]

In 1985, the grazing fee issue once again became an important natural resources topic in Congress. A loose, relatively peculiar coalition of environmental groups, cost-conscious Reagan officials, and some BLM and Forest Service officials formed in an effort to achieve higher fees, albeit for different reasons. Environmental groups wanted higher fees primarily to reduce the number of cattle grazing on the public lands. With the election of Reagan in 1980, a host of supporters of economic liberalism attained positions of importance. Primarily economists, especially within OMB, they sought to have the free market play a more important role in numerous policy areas, including grazing. The budget officials viewed the low fees as a subsidy and, with a large budget deficit, were looking to eliminate any unnecessary spending. An OMB official stated, "In all fairness to the taxpayer, these grazing fees are too low. . . . It has been our position for years that these fees, as well as other user fees, have been far below their . . . value." Bureau of Land Management and Forest Service officials favoring the increase viewed it as an opportunity to increase fees that were too low and to achieve some political independence from livestock interests.[64]

In addition to their demand for higher fees, environmental groups also wanted a greater environmentalist voice on the advisory boards and increased protection for critical riparian areas. Nevertheless, the main issue was the increase in fees and the reduction in livestock numbers. As the National Wildlife Federation stated: "We basically believe that the people who desire economic advantage from publicly owned resources should pay a fair price for them." The 1985 government fee was $1.37/AUM, and comparable private lands had a fee of $6.00/AUM or more. This low fee meant that BLM's grazing management program was typically in the red, with annual costs exceeding revenues from $30 to $50 million. Environmentalists claimed that no "single activity or combination of activities [has] contributed more toward altering the shape and texture of the land and wildlife that is dependent upon it" than livestock grazing in the West. By increasing fees, the number of livestock on these lands would be reduced. Even though herd sizes might be reduced only slightly, "The effect it *would* have," according to the Wilderness Society, "is to remove cattle and sheep from the more marginally productive lands. It would also increase the funds available for restoration and range improvements." Further, an editorial in *Outdoor Life* magazine proclaimed that "fish and wildlife's biggest enemy is the excessive livestock grazing being done on more than 200 million acres of rangelands managed by the Bureau of Land Management and the U.S. Forest Service."[65]

Environmental groups also increased attacks on the power of livestock interests in the West. They explored and cited stories of BLM or Forest Service officials who were fired because they had tried to implement programs unacceptable to livestock interests. They also stressed the close relationship of western range-management academics and livestock interests. The political clout of public-lands ranchers in the West remained high, even with the increased development and urbanization in the area. In key states, use of public grazing lands was very high: 88 percent of the livestock in Idaho grazed on public lands at some time during the year, 64 percent in Wyoming, and 63 percent in Arizona.[66]

In March 1985 the BLM and the Forest Service completed a four-year study in which twenty-two professional appraisers collected data for every county in the West that had rangeland. The report discovered that fees on private lands averaged $6.87/AUM, almost five times higher than public fees at the time. The fees on state lands ranged from $1.43/AUM to $14.00/AUM, and the fees charged by other federal agencies averaged $6.53/AUM. This study and the results it contained supplied further evidence to the forces that wanted to significantly increase grazing fees.[67]

Livestock interests did not stand idly by during this period. They made their positions known and began what they hoped would be another successful venture into grazing policy in Congress. The ranchers were quick to attack the new BLM/Forest Service report, stressing that public and private lands could not be compared. The public lands were inferior due to the numerous management expenses the rancher had to pay—for example, for

Table 6. Quality of Bureau of Land Management Rangelands, for Selected Years (%)

Year	Excellent	Good	Fair	Poor
1936	1.5	14.3	47.9	36.3
1966	2.2	16.7	51.6	29.5
1975	2.0	15.0	50.0	33.0
1984	5.0	31.0	42.0	18.0

Source: Council on Environmental Quality, *Environmental Quality 1984*, p. 259.

fences and water and for other users. The National Cattlemen's Association declared the work "a report that compares the present fee based on the value of forage with fees based on values other than forage."[68]

The ranchers also challenged the environmentalists efforts to link grazing fees with overgrazing. Environmentalists argued that overgrazing reduced the forage available for wildlife and increased erosion, which fouled water quality and generally hurt fish populations. They argued that low fees encouraged overgrazing. By raising the fees, some marginal grazing would be stopped and the higher fees would translate to more funds for range improvement. Wildlife supporters pointed to figures suggesting low productivity on BLM lands, and the decline of big game animals, which were outnumbered by livestock on BLM land 4.3 million to 1.5 million. Ranchers countered that the fee issue was separate from the issue of range condition. According to the National Cattlemen's Association, "The fee is an economic process and should be dealt with separately and not as a management tool." Because the ranchers had no discretion in setting stocking rates on these lands, how, asked ranchers, could fees and overgrazing be related? Lastly, livestock interests pointed to government statistics that indicated that the quality of the BLM lands were improving (see table 6). Although environmentalists acknowledged some improvement, they were quick to point out that 60 percent of the rangelands were still in unsatisfactory condition.[69]

Congress returned to the issue as well. The first major bill, introduced by Representative John Seiberling (D, Ohio), would have used a competitive bid system to determine grazing fees. This system was estimated to cost over $70 million more than the current system in the short term and was immediately attacked by livestock interests. Soon afterward a joint House-Senate discussion bill was circulated by Representatives Seiberling and Morris Udall (D, Ariz.) and Senators James McClure (R, Idaho) and Malcolm Wallop (R, Wyo.). These four members of Congress held the key institutional roles in the debate: chair of the Subcommittee on Public Lands, chair of the Committee on Interior and Insular Affairs, chair of the Committee on Energy and Natural Resources, and chair of the Subcommittee on Public Lands, Reserved Water, and Resource Conservation.[70]

The discussion bill was one that was not entirely acceptable to either environmentalists or livestock interests. It was hoped, however, that some compromise could be worked out. The four main topics addressed in the bill were: (1) restoration of overgrazed riparian areas and protection of these areas in the future, (2) alteration of the membership of advisory boards so that there would be less grazing representation and more environmental and wildlife representation, (3) inclusion of provisions to remove lands unsuitable for grazing from existing grazing districts, and (4) continuance of the current fee formula. In discussions with environmentalists and livestock interests, both sides could agree to the first two points, the riparian areas and the advisory boards. Conflict was intense on the other two issues, however. Environmentalists would not accept an extension of the current fee formula; the livestock interests were opposed to allowing bureaucrats to remove "unsuitable lands" from grazing.[71]

As discussion on the bill continued, it became clear that the grazing fee was the key issue. If a compromise could be worked out on this, it was likely the bill would be successful. No sign of compromise on the fee was apparent, however. Environmentalists and the OMB strongly opposed any extension of the existing fee formula. Livestock interests strongly favored extending the current fee formula. Senator Jake Garn (R, Utah), an ally of livestock interests, claimed that a "fee hike would 'destroy the family rancher's way of life and the spirit or even the existence of some Western communities.'" A bill also under consideration at the time, introduced by Representative Dick Cheney (R, Wyo.) and cosponsored by twenty-seven mostly western Republicans would make the current grazing-fee formula permanent but dealt with nothing else. Eventually, the discussion bill died, due to a stalemate between environmentalists and livestock interests on grazing fees. Because Congress did not pass new grazing-fee legislation and authorization for the PRIA fee formula expired at the end of 1985, the grazing fees for 1986 would be decided by the administration.[72]

As 1986 approached, environmentalists and livestock groups turned their attention from Congress to the executive branch. Because Congress was unable to enact a law, the fee for 1986 and beyond (until Congress did act) would be set by the administration. The National Wildlife Federation, among other environmental groups, wrote directly to the president, urging

Reagan to raise the fees to FMV. The livestock groups summoned the heavy artillery for their presidential lobbying. On December 13, 1985, Senator Paul Laxalt (R, Nev.), a close friend of the president's, delivered a letter from a group of primarily western, Republican senators urging him to freeze the grazing fee. The letter argued that the low fees had been capitalized into private property, and that the ranchers had to pay for management costs on the public lands; hence the low fees were justified. Ironically, among the group signing the letter to freeze the fees were nineteen senators who had voted for the Graham-Rudman-Hollings Act.[73]

Although the congressional mandate for the PRIA fee system expired on December 31, 1985, the BLM announced that it would use the system at least through the end of February 1986, at which time a new grazing season would begin. In late January 1986, the OMB abandoned its coalition with environmental groups in seeking FMV grazing fees. In a letter to President Reagan, OMB director James Miller wrote: "The issue of the appropriate level of grazing fees on federal rangelands is of great political sensitivity." He went on to recommend a one-year freeze on fees to "maintain the pressure on Congress to seek a permanent solution to this problem." Despite OMB studies estimating that grazing revenues only covered approximately 35 percent of spending on such programs and that the government may have lost $500 million over the last ten years by not charging FMV, the agency bowed to political pressure. By this time, twenty-eight senators and forty representatives had urged the president to extend the current fee formula for at least ten years.[74]

The issue of grazing fees did not capture headlines, but at least one major newspaper was aware of the likely use of double standards by the administration. A *Washington Post* editorial stated: "In a year when so many other forms of federal support are in jeopardy, there is no excuse for exempting these [grazing fees]. The president keeps saying that if Congress can't bring itself to make hard decisions, the least it could do is give him the power to make them instead. Here he already has the power. He should use it." He did use his power, but he used it to maintain the current fee system.[75]

On February 14, 1986, President Reagan issued Executive Order 12,548. The order extended the PRIA fee formula indefinitely but added a fee floor of $1.35/AUM. Reagan cited the advice of Agriculture Secretary Richard Lyng and Interior Secretary Donald Hodel, who claimed the president's order would "maintain [the] stability of the Western livestock industry." Environmental groups were quick to attack the decision. The Public Lands Institute called the decision an "outrage . . . an unfair subsidy to a tiny minority of livestock operators in the West." The Izaak Walton League stated, "We are disappointed at this shallow attempt to pursue Western political interests." The BLM tried to take a middle position, emphasizing the lack of a relation between grazing fees and overgrazing.[76]

Nine environmental groups filed suit against Interior Secretary Hodel and Agriculture Secretary Lyng on May 12, 1986, over the fees. The groups charged that the fees violated federal laws requiring the government to charge FMV for the use of public lands. A Wilderness Society representative stated, "The subsidized fee amounts to paying the ranchers to destroy the public's land. But we are confident that the courts will step in and put an end to the Administration's senseless subsidy of destruction." Several measures to overturn the executive order were introduced in Congress, including one bill mandating a fee increase to over $4.65/AUM that cleared the House Appropriations Subcommittee for Interior and Related Agencies, but none of the bills received serious consideration.[77]

Reagan's executive order and the PRIA grazing-fee formula still remain in effect, though serious reform efforts have continued into the 1990s. The House has three times passed significant increases in grazing fees and, as detailed in Chapter 1, the Clinton administration tried to increase the fees. Nevertheless, ranching interests remain staunch in their opposition, as do western senators, and have thus far been able to block any change in the grazing fee formula.

The main issue in the grazing policy regime since the passage of the FLPMA has been grazing fees. Preservationists sought to have the fees increased to fair market value, arguing that this would reduce the number of cattle grazing on the land and hence increase the quality of the land and the amount of forage available for wildlife. Preservationists had an unusual ally in this battle: economic liberals committed to the free market. They argued that the fees were too low and were not stimulating efficiency. The fees should be set by the market, not BLM bureaucrats. Technocratic utilitarians at the BLM were weak supporters of the fee increase because they would gain funds to increase management. They were not strong supporters, though, because they did not want to alienate livestock interests.

Ever since losing the battle to keep public-domain lands their free domain (lost in the Taylor Grazing Act), the livestock interests, supporters of interest-group liberalism, have dominated the setting of fees on Grazing Service/BLM lands. These fees were kept very low, below $.20/AUM, through the 1960s, and although the fees have increased in the 1970s and 1980s, the ranchers have been able to rely upon the institutionalization of interest-group liberalism to prevent the fees from increasing substantially.

These ranchers argued that the current low fee system best served the public interest by helping marginal ranchers stay in business. Interest-group liberalism still held a privileged position in grazing policy, a privilege built into the Taylor Grazing Act and one that has yet to be overcome by either the BLM or the preservationists.

Conclusion

Interest-group liberalism is the idea that has dominated the grazing policy regime. Based upon this idea, embedded within the state at the inception of the grazing policy regime, grazing policy was instituted with an extralegal property-rights system, low fees, and "home rule on the range." This initial system led to the development of the captured policy pattern that has characterized grazing policy since the 1930s.

Two significant challenges to the privilege of interest-group liberalism arose in the 1970s. First, technocratic utilitarians sought to make the BLM a more professional agency. With the help of preservationists, the FLPMA was passed and the BLM had an organic act. The new management directives in the act, and its preservationist sections, represented cracks in the privilege of interest-group liberalism. But not severe ones. Supporters of interest-group liberalism were able to gain special policies for grazing in the law. Furthermore, preservationists (and free-market economic liberals) have been unsuccessful in their efforts to increase grazing fees. Thus far, the approximately 16 percent of western ranchers who hold BLM grazing permits have been able to rely on the remaining privilege of interest-group liberalism to thwart this challenge.

Although interest-group liberalism remains privileged in grazing policy, there are signs that this privilege may be in the process of being dislodged. The success of technocratic utilitarians in the passage of the FLPMA and of preservationists in wilderness reviews of grazing lands indicate this. Furthermore, the passage of significant increases in grazing fees in the House and the Clinton administration's efforts to increase fees suggest that the institutionalization of interest-group liberalism and the captured grazing policy regime may not last much longer.

EXPLAINING THE PAST, SPECULATING ABOUT THE FUTURE

The puzzle that motivates this study is why policy patterns within the realm of public lands are so different from one another. I have traced these differences to the fact that in each of the three cases examined, an idea associated with the dominant thinking about state-society relationships at the time became embedded in government policy about the land-use issue in question. In hard-rock mining policy, the idea was economic liberalism; in forestry policy, it was technocratic utilitarianism; and in grazing policy, it was interest-group liberalism. These ideas gained a privileged place in their respective domains of state policy. Once embedded in government, they served to channel and constrain policy within a particular regime, establishing in the process a particular pattern of development. These results are summarized in table 7.

Table 7. Ideas, the State, and Policy Patterns in Public-Lands Politics

	Policy Regime		
	Mining	Forestry	Grazing
Privileged idea	Economic liberalism	Technocratic utilitarianism	Interest-group liberalism
State capacity	Nonexistent	Strong	Weak
Extralegal property rights	Yes	No	Yes
General policy pattern	Privatized	Professional	Captured

Although these ideas became institutionalized and privileged, they did not go unchallenged. As the cases demonstrate, societal groups and administrative actors supporting nonprivileged ideas have launched efforts to overcome the privileged ideas. Beginning in the 1960s, cracks in the foundations of privilege began to develop in each of the three policy regimes. In hard-rock mining policy, the Wilderness Act and land withdrawals have been significant successes for supporters of preservationism. In forestry policy, the Wilderness Act severely jolted the privilege of technocratic utilitarianism. This crack has been wedged open further by preservationists, who have succeeded in having the wilderness system greatly expanded in size, in having new regulations developed regarding clear cutting, in having the Forest Service decision making process opened to tremendous public participation, and in penetrating the Forest Service itself with the formation of the Association of Forest Service Employees for Environmental Ethics. The privatization challenge of the economic liberals, although unsuccessful, in part due to a coalition between the supporters of technocratic utilitarianism and preservationism, was nevertheless another significant challenge. In grazing policy, preservationists were successful in having wilderness review extended to BLM lands and in having a special management area created in the California desert. They also were among the leaders in the efforts to have grazing fees on BLM lands raised to free market value. The technocratic utilitarians were successful in achieving the passage of the FLPMA, which allowed them to develop management policies based upon multiple use and sustained yield and to undertake long-term land-use planning. Despite these challenges and cracks, the embedded idea has proved difficult to dislodge. In each of the policy regimes, the privileged idea, though tarnished, is still in place.

The most important effect of ideas on public-lands policy has been in shaping the initial state within the specific policy regimes. In all three of the areas examined, one idea played a central role in shaping the state, an influence that has continued to the present. The mining policy regime was shaped by economic liberalism. The prospectors and miners had created a private mining policy for the public lands based on mining codes and an extralegal property-rights system. When the federal government finally acted to establish a mining policy for these lands, it basically legitimized the existing private policy. The mining codes and extralegal property-rights system were recognized as valid, and the law established a relatively simple process for prospectors and miners to file claims and obtain patents to land through the federal government. No new agency was established to oversee this program. It was simply another, albeit specialized, part of the overall public-lands policy of the time: transfer these lands to the private sector to achieve the public interest. Indeed, it was a piece of the prevalent idea of the time for government to help "the release of energy" within society. Economic liberalism became embedded in the mining policy regime and served to shape the privatized policy pattern that still exists today, most obviously in the continued reliance on the 1872 Mining Law as the core of hard-rock mining policy on the public lands.

In the forestry case, technocratic utilitarianism was the central idea: forest resources should be managed by experts for the greatest good for the greatest number for the longest time. The Forest Service, under the guidance of Pinchot, was organized around this idea. The agency gained authority over the national forests and employed various methods to insulate itself from political control and develop a degree of autonomy (e.g., civil service requirements, publicity efforts, control of forest-reserve receipts). The Forest Service also adopted an internal philosophy based on this embedded idea, later articulated as multiple use and sustained yield, which continues to be the dominant philosophy of the agency. All of this, based on the privileged idea of technocratic utilitarianism, helped establish the professional policy pattern in forestry.

In the grazing policy regime, interest-group liberalism became embedded in the state. Dominating the debate over the Taylor Grazing Act was whether to establish a federal grazing system that continued the status quo favoring the existing livestock operations—the legitimation of the existing system. The influence of interest-group liberalism is reflected in the weak Division of Grazing/Grazing Service, which was captured by livestock interests; the central role of the local grazing advisory boards, which allowed the ranchers to determine who would receive permits and how much grazing those permits would allow; and the inclusion of the existing extralegal property rights in the Taylor Grazing Act. From this beginning, the privileged idea of interest-group liberalism helped forge the captured policy pattern in the grazing policy regime. As the case study on grazing fees demonstrates, the part of the state responsible for dealing with grazing policy remains greatly shaped by this idea.

In each policy regime the embedded idea has been challenged recently by the supporters of other major ideas. The embedded idea has thus far survived as the privileged idea in each policy regime, but the foundations of privilege have been eroded. Significant cracks have appeared in the institutionalization of the privileged ideas. These cracks have developed in response to both societal and state changes, the most obvious such change being the rise of environmentalism in society at large. The rise of this movement, traced to factors such as increasing affluence and increasing population, helped fuel the preservationist challenges in each policy regime. Similarly, poor economic conditions and the seeming failure of past economic theories led to the resurgent popularity of free-market economics, which helped fuel the privatization initiative and the desire for increased commodity access to the public lands. These specific societal changes may indeed be part of a larger restructuring of state-society relationships. As more interests have become organized and active in the national political world, as privileged ideas have been challenged and dislodged, we may be entering a new political era, one dominated by pluralism. The chief advantage of such a system is the increased representation of societal interests; the chief drawback is the potential for gridlock as groups on either side of issues are powerful enough to prevent significant policy decisions of any kind.

An important lesson to be drawn from this study is that even within a limited area of state concern—public-lands policy—governance takes on very different capabilities and characteristics. A virtual absence of the state marks the mining policy regime. For lands open to free access, minerals are simply transferred to the private sector with only a limited clerkship role for the federal government. This lack of a state presence can be traced back to the period when the 1872 Mining Law was passed, the establishment of federal mining policy on the public lands. At this time, there was little administrative state of any kind at the national level. The extralegal property-rights system adopted as part of this policy became the core of this privatized policy regime, institutionalized in the 1872 mining system. The chief change has been the removal of land from free access, rather than a fundamental altering of the core policy.

In the forestry policy regime, the Forest Service demonstrated strong state capacity from its inception. It had strong leadership (Gifford Pinchot), was relatively insulated from political pressure (through the civil service and a nonpolitically appointed chief), was well funded and well staffed, was professional, and enjoyed a good relationship with both Congress and the president during its formative years. During these first few years of existence, the Forest Service even demonstrated characteristics of state autonomy. This early start bode well for the future of the Forest Service, and into the 1960s it was cited as one of the most effective federal agencies.

Quite a different picture of the state is found in the grazing policy regime. The Division of Grazing/Grazing Service/BLM demonstrates the existence of a chronically weak state agency, with the agency often cited as a classic example of a captured agency. The grazing state was born weak for a number of reasons: the lack of a profession of range management, low funding and low staffing, the existence of local grazing advisory boards, numerous reorganizations in its early history, a radically decentralized administrative structure, conflict with the Forest Service over control of the grazing program, the political nature of its employees, attacks on the new agency from within the Interior Department, and the incorporation of an extralegal property-rights system weighed strongly in favor of permittees rather than government. Some of these characteristics have changed; most notably the BLM is now professionalized. Yet despite these gains, state capacity in the grazing policy regime is still weak, haunted by its past.

I should stress at this point that in the hard-rock mining and grazing policy regimes that state acceptance, incorporation, and legitimation of extralegal property-rights systems were of crucial importance, especially in the institutionalization of the economic and interest-group liberalism perspectives. By adopting these systems, the state was accepting the status quo that had been in operation prior to federal policy. In both cases, this status quo was based on developing the resources to further private interests, which would in turn further the public interest. In both of these cases the respective property-rights systems were the core around which the new policies (the 1872 Mining Law, the Taylor Grazing Act) were constructed. These systems restricted management flexibility in both policy regimes and continue to do so today. In efforts to alter access to minerals or change grazing fees on the public lands, the issue of a violation of rights is a formidable barrier.

There are also certain commonalities of the state in the more general public-lands policy arena that affected policy in each of the distinct policy regimes. The overall fragmentation of the state in the public-lands arena had consequences in each regime. The fragmented mining policy regime worked against the development of a coherent and coordinated strategic (and general) minerals policy. Repeated efforts at bureaucratic reorganization stand as testimony to the fragmentation in this policy regime. In forestry, the competition between the Interior Department and the Agriculture Department helped spur the development and spread of administrative wilderness areas. And even earlier, the placement of the Forest Service and the national forests in the Agriculture rather than the Interior Department served to fundamentally fragment the state in public lands policy from the outset. This fragmentation contributed both to the delay in the passage of the Taylor Grazing Act and to its ultimate weakness, because the Interior and Agriculture Departments battled for more than a decade to gain control of the program. When it became clear that the new policy regime would be based in the Department of the Interior, the Forest Service urged the president to veto the Taylor Grazing Act. In addition, this fragmented state in general was a fertile ground for various ideas to take hold.

Little attention has been paid to the courts in this study, which played a significant role in these policy regimes, often in specifying the details of federal laws and programs. Indeed, significant cases in these policy regimes helped determine the shape and powers of the developing American state. In the mining policy regime, two important Supreme Court cases involved the lead-leasing program. In *United States v. Gratiot* (1840), the Court held that the Constitution grants Congress the power to lease, as well as sell, public lands. The decision established a judicial foundation for the entire public-lands management programs that followed. In *United States v. Gear* (1845), the Court ruled that unauthorized mining on the public lands is an actionable trespass. Debate over the public lands also helped define the powers of the nascent administrative state. The Supreme Court cases *United States v. Grimaud* (1911) and *Light v. United States* (1911) legitimized the discretionary use of executive power (regarding the Forest Service's authority to collect grazing fees for the use of national forest lands without a specific act from Congress authorizing it to do so). In later policy debates, for example, on wilderness, court decisions interpreting the law had major policy implications. Despite these significant decisions, the courts did not play a major role in altering policies determined by the legislative and executive branches in these policy regimes.[1]

Public-Lands Politics Today

As I have sought to demonstrate throughout this book, hard-rock mining, forestry, and grazing policy on the public lands have been shaped by the ideas institutionalized at each policy regime's inception and by the subsequent policy patterns that have developed. These institutionalized ideas, though, have not been stagnant. In this section, I examine the current tensions within economic liberalism, technocratic utilitarianism, and preservationism as well as examine the current status of mining, forestry, and grazing policies.

Since the policy failures of the Sagebrush Rebellion and privatization in the late 1970s and early 1980s, economic liberalism has been dormant in the public-lands policy arena. Energies that might have been used to further economic liberalism approaches have been channeled into the interest-group liberalism approach, the wise use movement. The founding of this movement can be traced to the August 1988 Multiple Use Strategy Conference. This gathering of individuals and firms that used the public lands sought to counter an environmental movement that, in its view, was closing the public lands to multiple use, especially commodities development and motorized recreation. At the conference, an agenda of twenty-five goals was adopted. Many of these agenda items refer to public-lands issues, such as the extension of the 1872 Mining Law to all national parks and wilderness areas; "Passage of the Global Warming Prevention Act to convert in a systematic manner all decaying and oxygen-using forest growth on the National Forests into young stands of oxygen-producing, carbon dioxide-absorbing trees"; the identification and preservation "for commodity use those timberlands suitable for sustained yield timber growth" and the prevention of any timber sale being identified as "below cost"; the creation of a National Rangeland Grazing System that defines grazing as the first and foremost use of public rangelands; the recognition of private possessory rights to mining claims, timber contracts, and grazing permits on public lands; and the reevaluation of existing wilderness, with a recommendation that only twenty million acres (of a total of one hundred million) be retained in the current system.[2]

The wise use movement has grown rapidly since 1988. It is centered in the rural West, in those communities that depend on mining, forestry, and ranching. In addition, wise use has become the grass roots in support of commodity development on the public lands. The movement consists of a variety of loosely affiliated groups, such as Alliance for America, National Inholders Association, and People for the West! Among the business groups that are members or financial backers of wise use groups are the American Mining Congress, Boise Cascade, Georgia-Pacific, Louisiana-Pacific, and the National Cattlemen's Association. Thus far, the movement's chief role has been to help organize opposition to new environmental policies affecting the public lands, such as reforming the 1872 Mining Law and grazing reform, although the movement's agenda has become more activist since the Republicans gained majority status in Congress. As this brief discussion makes clear, the wise use movement focuses on favorable conditions for commodity use on the public lands, not privatization. Hence, it is best described as furthering the idea of interest-group liberalism rather than economic liberalism.[3]

Supporters of technocratic utilitarianism have increasingly focused on two new policy ideas consistent with its general themes, sustainable development and ecosystem management. There have also been recent movements within the bastion of technocratic utilitarianism, the Forest Service, that are working to undermine the privilege of technocratic utilitarianism. These challenges include the rise of the Association of Forest Service Employees for Environmental Ethics (AFSEEE) and the increasing diversification of Forest Service employees.

Since technocratic utilitarianism was institutionalized in forestry policy in the early 1900s, the dominant mode of practice on the land has been multiple use and sustained yield. That is, the national forests should be managed for a number of resources, chiefly fish and wildlife, forage, recreation, timber, and watershed protection, and these resources should be managed in such a way that they will be available to future generations. This policy idea was formally incorporated into Forest Service policy in the Multiple Use and Sustained Yield Act in 1960 and in the BLM in the Federal Land Policy and Management Act in 1976. More recently, however, new ideas have sought to reflect increasing scientific knowledge of how ecosystems work and growing societal support for environmentalism. Two main ideas that have arisen are sustainable development and ecosystem management.

Sustainable development—as an idea and a concept—gained increasing usage following the release of *Our Common Future* in 1987, a World Commission on Environment and Development report completed under the leadership of Norwegian prime minister Brundtland. In a recent study of the concept in the United States, *Choosing a Sustainable Future*, sustainable development is defined as "broad-based economic progress accomplished in a manner that protects and restores the quality of the natural environment, improves the quality of life for individuals, and broadens the prospects of future generations." In many ways, sustainable development is simply a new version of Gifford Pinchot's dictum "the greatest good to the greatest number for the longest time"; it is essentially multiple use and sustained yield applied to all human actions.[4]

Ecosystem management is a more specific idea geared to land management. One definition of ecosystem management is "any land-management system that seeks to protect viable populations of all native species, perpetuate natural-disturbance regimes on the regional scale, adopt a planning timeline of centuries, and allow human use at levels that do not result in long-term ecological degradation." The key to ecosystem management, based on a better understanding of how ecosystems work, is the need to manage in terms of uncertainty, change, risk, variability, and complexity. Under this new approach, the resource is viewed holistically and sustainability is viewed dynamically; under the old approach—multiple use and sustained yield—individual resources were managed and sustainability was viewed statically.[5]

In the 1990s, the Forest Service has attempted to translate ecosystem management into practice, focusing its efforts on the Pacific Northwest. The first effort was stimulated by the spotted owl debate in Washington, Oregon, and California west of the Cascades. More recently, the Eastside Ecosystem Management Project, encompassing thirty million acres of federal land in eastern Washington and Oregon, all of Idaho, and parts of Montana, Wyoming, Utah, and Nevada, has been established. In both planning efforts, Forest Service scientists are running the processes (covering BLM and Forest Service land), ones so complex that only the most dedicated citizen can maintain active involvement.[6]

Both sustainable development and ecosystem management are ideas that fit well under the technocratic utilitarianism umbrella. Both rely on experts to make complex management decisions to further the public interest. These experts must determine what sustainability means and what actions do and do not meet these criteria; they must determine how ecosystems function and how human management of these various ecosystems can succeed.

The appointment of Jack Ward Thomas as chief of the Forest Service in late 1993 is somewhat ambiguous in relationship to the place of technocratic utilitarianism in the Forest Service. Many have pointed to his appointment as a sign of change: he is the first biologist to hold the top position, he was the chief government scientist in the spotted owl case, and his selection violated Forest Service tradition in that he was not part of the top-level of agency officials from whom the chief has always been selected. On the other hand, Ward has a number of characteristics that link him to technocratic utilitarianism: his Ph.D. is in forestry, he is a firm believer in ecosystem management, he thinks clear cutting is an appropriate approach to forest management, and in a recent interview he supported laws that gave "professionals more latitude" to manage the land.[7]

The creation and rise of AFSEEE is less ambiguous. This group, founded in 1989, "seeks to forge a socially responsible value system for the US Forest Service based on a land ethic which ensures ecologically and economically sustainable management." By 1994, the group had more than ten thousand members, including twelve hundred current, former, or retired Forest Service employees. Among the goals of AFSEEE are the protection of "all remaining old-growth forests and roadless areas on public lands," support for a noncommodity orientation, and a focus on preserving ecological integrity for the Forest Service. In general, the group is opposed to the privileged position of technocratic utilitarianism within forestry policy; it supports following the laws passed by Congress (rather than increased agency discretion) and preservationism.[8]

Recent empirical studies of Forest Service employees suggest the erosion of the dominance of technocratic utilitarianism within the agency is more widespread than just within the AFSEEE membership. One study of district rangers and forest supervisors found that since 1981 people in these positions have become less inclined to support commodity development on the national forests, more supportive of recreation, and significantly more concerned about the environment. A related study suggests that the Forest Service workforce is becoming more diverse in terms of gender and professions (e.g., fewer foresters and engineers, more biologists, ecologists, and recreation planners), and that women and those in the nontraditional professions are more supportive of environmental concerns and less supportive of commodity development than men and those in traditional Forest Service professions.[9]

As the above discussion indicates, there is a real tension within the resource management professions. This area has long been dominated by the idea of technocratic utilitarianism, and sustainable development and ecosystem management suggest the continued dominance of this idea. But the rise of AFSEEE and Forest Service employee diversification suggest that other ideas, namely, preservationism, are gaining significant ground in the resource-management professions. As the strength of technocratic utilitarianism is undermined within these professions, the overall importance of the idea in public-lands politics will be weakened.

In the fall of 1994, preservationists celebrated their greatest success in fifteen years with the passage of the California Desert Protection Act. After an eight-year legislative battle, the law established nearly eight million acres of new wilderness on BLM and National Park Service land in southern California. This was the single greatest preservation law since the passage of the Alaska Lands Act designated nearly sixty million acres of wilderness in 1980. Supporters of preservation, however, were undergoing internal disputes at the same time, most vividly demonstrated on the Montana wilderness issue.

The biggest change within preservationism has been the rise of radical environmentalism, which has challenged the foundations and the political strategies of mainstream preservationism. At the core of preservationism is protecting certain land due to its recreational, scenic, and spiritual values for humans. This goal was achieved through the creation of national parks and the designation of wilderness areas. Preservationists did not question the basic structure of the economy and society in the United States; they sought to achieve their goals within what existed. Each of these tenants were rejected by many of the radical environmental groups, most visibly by Earth First! These new groups sought to preserve land due to the values this land had for other species and the functioning of ecosystems. They too sought the protection of these lands primarily through wilderness designation, but they favored much broader proposals, seeking the protection of up to 50 percent of North America as wild lands. They argued for restoration, and some made use of ecotage (ecological sabotage) or monkeywrenching to prevent the destruction of wild lands. These radical environmentalists challenged the fundamental structure of society. Many support deep ecology, based on ecocentrism, which holds that decisions affecting nature should be based not only on how they affect humans but also on how they affect nature. That is, other parts of the planet or ecosystem have the right to survive, to be left alone.[10]

This rise of deep ecology and radical environmentalism has led to increased tension within preservationism. Most mainstream preservation groups have attempted to distance themselves from radical environmentalists, especially when monkeywrenching has been involved. They have sought to follow their traditional course of working through the system to achieve increased land protection. The existence of the radical groups, however, has opened the preservationists up to criticism from within: why are they settling for so little protected land? why have they been so willing to compromise? why don't they focus more attention on protecting lands rich in biological diversity? why haven't they questioned the foundations of American society?

Although much of this debate has been internal to preservationism, it has occasionally spilled out into the policy arena. This is perhaps best illustrated in the debate over Montana wilderness. Efforts have been under way to designate more wilderness in Montana since the early 1980s. In 1988, Congress passed a bill establishing 1.4 million acres of new wilderness in the state, only to have it vetoed by outgoing President Reagan. Over the next few years, preservationists battled with commodity-oriented interests over the shape of another Montana wilderness bill. Then, a new group—Alliance for the Wild Rockies—got involved, making a far grander proposal: protecting more than fifteen million acres in five core ecosystems in Idaho, Montana, Oregon, Washington, and Wyoming, with the protected lands connected by corridors for wildlife movement. This set off a major political debate within the preservation community: should preservationists support a more practical bill just dealing with Montana, setting aside somewhat less than two million acres (as favored by Montana representative Pat Williams), or should they support the idealistic Northern Rockies Ecosystem Protection Act (NREPA), which many viewed as politically impossible?

What has resulted is something of a civil war. Mainstream national environmental groups opposed NREPA, whereas many grass-roots groups supported it. Within the Sierra Club, for instance, different chapters took different stands on the bill. The national leadership was against it, even lobbying members of Congress to oppose its introduction. Many of its chapters, however, actively supported NREPA, accusing the national organization of being coopted by Washington politics. Finally, in December 1993 the Sierra Club held a meeting in Montana to try to return harmony to the ranks. It adopted a multiple-track strategy, allowing the Sierra Club to support both NREPA and the Williams bill. Meanwhile, grass-roots activists have kept up the campaign for NREPA. It was introduced in the House in July 1993, gained sixty-four cosponsors, and was the subject of hearings in April and May 1994.[11]

So preservationism, like technocratic utilitarianism, is suffering from severe internal tensions. Although more than one hundred million acres of public lands have been designated as wilderness since 1964, a testament to the tremendous policy successes of preservationist supporters, the mainstream groups that have embodied this idea are under significant challenge from more radical groups. These more radical groups may have any number of effects on preservationism: they might shift the entire idea in a more radical direction; they might tar the entire idea, making any preservation gains more difficult; or they might have no effect at all.

The state of mining, forestry, and grazing policy on the public lands is just as much in flux as the ideas discussed above. Congress started to focus seriously on reforming the 1872 Mining Law in the late 1980s. In 1990, the House included a one-year moratorium on the granting of mining patents on federal lands as part of its fiscal year 1991 Interior Appropriations bill. The Senate had considered such a proposal, but it was defeated forty-eight to fifty. Due to this Senate opposition, the moratorium was dropped in conference. The next year the House again passed a moratorium, the Senate rejected it (forty-six to forty-seven), and the moratorium was dropped in conference. This time, though, all parties agreed that reform of the 1872 Mining Law would be seriously considered the next year in the relevant committees.[12]

A reform bill did make it out of the House Interior and Insular Affairs Committee in 1992, but Congress adjourned before the bill could be fully debated and voted on. The main provisions of the bill were an 8 percent royalty on hard-rock mineral receipts, an end to patenting of land (i.e., the government would maintain ownership of all mining lands), an annual rental fee of twenty-five dollars for claims, direction to the BLM and Forest Service to review their lands and withdraw lands not suitable for mining, and the creation of a reclamation fund for abandoned mines. A similar reform bill did not make it out of the Senate Energy and Natural Resources Committee. Minor reform occurred, however, through the Interior Appropriations bill. Senator Harry Reid's (D, Nev.) proposal, offered as an amendment, was approved. It required miners to pay fair market value for land they wished to patent, land to revert to the government if it were not mined, and strict compliance with environmental and reclamation laws. This reform package was dropped in conference because opponents thought the reforms were not significant enough and would hurt the chances of real reform. Conferees did accept a provision, which became law, requiring a new annual fee of one hundred dollars miners would pay to keep their claims active. This replaced the required one hundred dollars' worth of mining work to keep a claim active, which often scarred the land for no real purpose.[13]

In early 1993, the Clinton administration attempted to reform the 1872 Mining Law through the budget process by including a 12.5 percent royalty on hard-rock minerals in its budget proposal. Clinton dropped the proposal, along with efforts to increase grazing fees and eliminate below-cost timber harvests, in the face of strong pressure from western members of Congress. Both the House and the Senate passed mining reform legislation in 1993, but the bills were quite different. The House bill was basically the same as the one that passed in 1992. The Senate bill sought less far-reaching change: a 2 percent royalty on net value of the minerals, retention of patenting but charging market value for the surface land, and only requiring mining operations to meet widely varying state environmental and reclamation standards. The conference committee worked on and off beginning in June 1994, but by late September efforts at reaching an agreement collapsed as both representatives and senators felt they could compromise no more.[14]

The future of mining reform has been cast into further doubt with the Republican victory in November 1994. Promining legislators from Alaska are now chairs of the natural resource committees in both the House and the Senate, making congressional action to reform the 1872 Mining Law less likely. Environmentalists, however, are likely to keep up the pressure for reform, focusing their efforts on budgetary arguments.

The complexities of public-lands forestry policy in the 1990s are best illustrated in the spotted owl dispute in the Pacific Northwest. On this issue, supporters of interest-group liberalism sought to have the Forest Service continue to "get out the cut" in order to keep jobs and profits intact, supporters of preservationism sought to have old growth forests protected from logging, and supporters of technocratic utilitarianism—namely, many Forest Service employees—sought to maintain expert control over forest management on the national forests. The issue had its beginnings in 1972, when biologists studying the owl discovered its reliance on old growth forests, a habitat in significant decline due to timber harvesting. By 1978, the Oregon Endangered Species Task Force, consisting of representatives of the BLM, Forest Service, Fish and Wildlife Service, and state agencies, had agreed on a management plan to protect the spotted owl on federal forest lands in Oregon. This occurred even though the owl was only listed as threatened under an Oregon program, not the federal Endangered Species Act.[15]

As the Forest Service and BLM sought to manage to protect the owls, they came under increasing pressure from environmentalists and timber interests. Some environmentalists, thinking that the plans didn't do enough to protect the owls, filed administrative appeals challenging the plans due to the lack of an environmental impact statement (EIS) as required by the National Environmental Policy Act (NEPA). Both the Forest Service and BLM administrative appeals were rejected, but they laid the foundations for future legal appeals concerning the spotted owl. As owl management began to lead to reduced timber harvests, the timber industry began to question the needs of the owl, stressing the reduced supply of timber, increased costs to consumers, and loss of jobs. The federal agencies tried to chart a middle course, with a central concern being to prevent the spotted owl from being listed as threatened or endangered, since this would limit their management flexibility.[16]

The Forest Service dealt with the spotted owl through the forest planning process, based on National Forest Management Act (NFMA) language requiring "diversity of plant and animal communities." As this planning process continued, the timber industry became increasingly vocal in its criticism, arguing that too much timber was going to be sacrificed for the spotted owl. Environmental groups were also not pleased, and in 1984 they began a legal challenge of the Pacific Northwest Regional Guide, a document directing forest planning in the region, arguing that its provisions to protect the spotted owl were insufficient. Deputy Assistant Secretary of Agriculture Douglas MacCleery ruled that the guide needed a supplemental EIS (SEIS) on owl management, though he denied the environmentalists request to stop all timber harvesting in owl habitat until the SEIS was completed.[17]

The draft SEIS came under intense criticism from both sides; timber interests argued that too much timber would be lost to owl reserves, environmentalists argued that—based on biological evidence—the plan did not sufficiently protect the owl. Revisions to the draft by the Forest Service based on this input satisfied neither camp. Both environmentalists and timber interests filed administrative appeals over the final SEIS; both were denied by the Agriculture Department. Both sides then filed lawsuits. In response, federal judge William Dwyer issued a temporary restraining order blocking all timber sales in owl habitat in March 1989. As the Forest Service prepared the final SEIS, the Fish and Wildlife Service (FWS), responding to an environmental group petition, studied the spotted owl for listing as a threatened or endangered species. At first, the FWS rejected the petition, concluding that management efforts under way would prevent the spotted owl from becoming endangered. This led to environmental groups challenging the decision in the courts in May 1988. The judge ruled the FWS's decision was "arbitrary and capricious" and ordered the agency to reconsider its decision in November 1988.[18]

These two court cases moved the policy-making process outside of the executive branch and led to wholesale, national controversy. Indeed, the issue even became part of the 1992 presidential campaign. Meanwhile, the FWS proposed in 1989 that the owl should be listed as threatened, a decision made final in 1990. In response to the freeze on timber sales, an Oregon timber summit was held in 1989 and Senator Mark Hatfield (R, Ore.) guided an appropriations rider through Congress in 1989 to allow some timber sales to go forward. The federal Interagency Scientific Committee (ISC) was convened in 1990 to analyze the spotted owl information again. It recommended setting aside 8.4 million acres of forest land, with a reduction of logging levels in Oregon and Washington of 30 to 40 percent. By this time, the Bush administration was getting involved. It bypassed the ISC report, favoring an approach that allowed for more timber sales and less land set aside for the owl. When the Forest Service failed to adopt a plan to sufficiently protect the spotted owl, environmentalists went back to court to block timber sales based on the NFMA. The temporary injunction was made permanent until a sufficient protection plan was in place.[19]

Environmentalists also used the courts to force the FWS to designate critical habitat for the spotted owl under the Endangered Species Act. In 1991, the FWS proposed critical habitat of 11.6 million acres (of not already protected land). This acreage required to protect the owls just kept rising, from 87,000 acres in 1978 to 400,000 acres in 1981 to 2.2 million acres in 1985. The Bush administration invoked the God Squad (a review board consisting of cabinet secretaries and agency administrators) in 1991 to exempt timber sales on BLM lands from the Endangered Species Act. The God Squad voted to exempt the sales, but the decision was challenged by environmentalists in court and only applied to minimal acreage. Bill Clinton took over in 1993 and called a forest conference in April 1993 to deal with the issue. Following this conference, Clinton directed his administration to develop a viable, balanced plan. The administration proposal— Option 9—came under attack from environmentalists and timber interests even before it was made public. Despite this pressure from both sides, Option 9 was adopted and presented to the court in April 1994. In June, Judge Dwyer removed the injunction against timber sales on federal lands, and in December he ruled that the plan met the legal requirements of the NFMA and NEPA.[20]

In addition to conflicts over the spotted owl, the Forest Service is also under attack over harvesting levels and clear cutting. Preservationists attack many of the agency's sales as below cost (i.e., they cost the Forest Service more than they earn). Even some Forest Service employees, supporters of technocratic utilitarianism, claim the agency is overly committed to supplying timber, leading it to exceed sustained yield and violate environmental laws. In 1991, the regional supervisor for the northern Rockies was essentially fired for refusing to reach his assigned timber cut, which he claimed could not be done while practicing good forestry and following existing laws.

As discussed in detail in Chapter 1, grazing-reform efforts have proved to be as difficult to achieve as mining reforms. This issue made it out of conference committee in 1993, only to die in a Senate filibuster. As is the case in mining and forestry policy, grazing policy has become more open and more conflictual; the privileged idea of interest-group liberalism is close to being dislodged. Nevertheless, the captured policy pattern in grazing continues. In the wake of the November 1994 elections, this policy pattern is likely to continue, though the conflict level may rise if Secretary of the Interior Babbitt continues to make executive-branch policy or if environmentalists turn to the courts as they did in the Pacific Northwest.

The internal tensions within the main ideas regarding public-lands management and the cracks in the foundations of privilege for the embedded ideas in each public-lands policy regime that have developed have generally made for more open, competitive politics. Once the privileged idea seemed vulnerable, it led to increased challenges, increased successes, and a further weakening of the embedded idea in the state. Overall, public-lands politics is more openly competitive now than it has been since the late 1800s, and it is likely to remain so. All of these changes suggest a further erosion of, indeed, perhaps the dislodging of, the privileged ideas in these policy regimes.

Despite those significant and wide-ranging efforts at change in each policy regime, the embedded ideas and interests tied to them, though weakened, are still in place. The Clinton administration's efforts to make fundamental changes in public-lands politics have been almost wholly unsuccessful. After failing in Congress, Babbitt has been hard pressed to make much progress administratively on grazing reform; mining and forestry reforms are hung up in Congress. So, even though the combination of cracks in the foundations of privilege in the three policy regimes examined and internal tension within the dominant policy ideas suggests a period of potentially dramatic policy change, the embedded ideas will not be displaced easily.

The Future of Public Lands and Environmental Politics

What do these findings suggest about the future of public lands and environmental politics in general? Perhaps of central concern are institutions and institutional change. Some advocates of environmental protection favor the creation of new institutions, arguing that the Forest Service and BLM are incapable of serious reform to address environmental problems. Some of these preservationists seek the creation of a new biodiversity corps (rather than trust this responsibility to the existing Fish and Wildlife Service), a new wildlands management agency (rather than have the existing agencies continue to manage wilderness), a new forest management agency to replace the Forest Service, and in the FLPMA debate favored the creation of a new land-management agency to replace the BLM. Those advocating such new institutions are at least partially acknowledging one of the chief lessons of this study: once an institution is firmly established and a privileged idea embedded within it, that institution is very difficult to change. New agencies would bypass the existing agencies and policy patterns that are based on ideas and serve interests not supported by these environmentalists. In creating a new institution, the hope is that preservationism could become the privileged idea.[21]

The creation of such new institutions is problematic in at least two ways. First, a new agency often is designed based on compromise with opponents of the program it will implement. That is, the bureaucracy is not designed to be as successful as possible because it is established in a political context. This was made clear in the recent efforts to create the National Biological Survey (NBS) and to elevate the EPA to a cabinet department, when opponents sought to encumber the institutions with requirements supporters did not favor. The House passed a bill in 1993 authorizing the NBS, but only after strong opposition from conservatives who feared that the NBS would become a new regulatory agency and infringe on private property rights. Successful amendments require written permission before the NBS can enter private property and prohibit volunteers from collecting data for the study (due to fears that they will have an environmental bias). The EPA cabinet legislation stalled in the House over debate as to whether the agency should be required to weigh the costs of proposed regulations against potential benefits. The second problem is one of fiscal reality. During a period framed by a large budget deficit and significant public opposition to more government and more taxes, it would be extremely difficult to create new agencies.[22]

Given the difficulties of creating new institutions, what of changing existing ones? These difficulties have been the substance of this book. Although the embedded ideas and the interests supporting them in these three policy regimes are being challenged and show signs of losing their privileged position, it must be stressed that the ideas have been guiding policy from 60 to 120 years. They have also been under significant challenge since the early 1960s, yet despite these challenges, the privileged ideas remain in place. It is the friction or stickiness of ideas that has slowed the change in policy regimes despite these societal challenges. It is the friction of these embedded ideas that helps to explain why the Mining Law of 1872 still determines hard-rock mining policy in spite of massive environmental opposition, why forestry policy, once a relatively calm policy regime, has become among the most controversial policy arenas in the nation, and why grazing fees have remained so low in spite of challenges from OMB and environmentalists. In a sense, public lands and environmental politics are seeing the gridlock of pluralist politics that is affecting politics in general. With the intensification of interest-group politics since the 1960s, it is extremely difficult to achieve significant policy change of any kind. Each group is strong enough to prevent significant change, even though they are not strong enough to push such change through the system.

In environmental politics, environmental groups are balanced by commodity and business groups. The status quo is generally maintained, with occasional dramatic shifts (e.g., the Wilderness Act). The privileged idea embedded within a policy regime is further protected by this pluralistic gridlock. In public-lands politics this favors commodity interests. The chief difference with gridlock in environmental politics and other political arenas is the stakes. We are in the midst of an environmental crisis, and if we can't improve our institutions and alter our policy patterns, we will pay the price in extinctions and irreparable harm to basic ecological systems. What is needed, then, is a fundamental change in state and society.

At the state level, we are in a period similar to the turn of the century or the Depression. The state doesn't really have the capacity to deal with the current set of problems—especially environmental problems—with which it is confronted. We are in need of a developmental surge like the one that took place in the early 1900s, transforming a state based on courts and parties into one with a significant administrative capacity, and following the Depression and World War II, when the administrative state expanded yet again to deal with broad economic policy and fighting a global war. Such a developmental surge could help free environmental politics from old policy patterns and institutions and from the current pluralist gridlock. Such a developmental surge could come on a number of fronts, such as increased state control over private property on environmental matters, increased state capacity in environmental policy, or moving to integrate interests into policy making in a corporatist model (rather than the current conflictual approach). Without such fundamental change, it is unlikely that our institutions will be sufficient to help us through the environmental crisis.

It is unlikely that any developmental surge will occur unless the environmental crisis becomes more focused or society demands such changes. Current environmental problems, such as the loss of biodiversity or global climate change, are not focused enough to bring about such state development. Within society, such demands are not likely under the current world view that humans have dominion over the natural world. If, however, a new world view reintegrating humanity into nature takes root and flourishes, society is likely to make demands on the state leading to a developmental surge supplying us with the institutions that could deal with our growing environmental problems.

NOTES

Chapter One

1. Taylor, "President Will Not Use Budget to Rewrite Land-Use Laws," pp. 833–34.

2. Hackett, "Wallop Proposes Grazing Fee Increase," p. A1.

3. Camia, "Administration Aims to Increase Grazing Fees," p. 2223; Krauss, "Clinton Planning to Increase Fees," p. A1.

4. Schneider, "Senate Hands Clinton Setback on Grazing Fee," p. A21; Schneider, "House and Senate Agree to Raise Fees for Grazing," p. A27.

5. Egan, "Wingtip 'Cowboys' in Last Stand," p. A1.

6. Camia, "Filibuster Ends," pp. 3112–13; Cushman, "Grazing Fees Cut from Senate Bill," p. A20.

7. Cushman, "Consensus Approach on Land Use," p. 8.

8. Davis, "Babbitt Cedes Grazing Reform," p. 6.

9. Council on Environmental Quality, *Environmental Quality*, p. 308; U.S. Department of the Interior, Bureau of Land Management, *Public Land Statistics 1991*, pp. 5–6.

10. A policy pattern refers to a stable set of tendencies or characteristics that describe how policy is made, debated, and implemented within a specific substantive policy area. This pattern also describes the distribution of power within the policy regime, both between state and society and among societal groups, and how these actors interact within the policy regime.

11. Theodore Lowi developed the first policy typology arguing that "a political relationship is determined by the type of policy at stake, so that for every type of policy there is likely to be a distinctive type of political relationship." He sketches three different types of policies: distributive, regulatory, and redistributive. Specific policy types within these larger categories are said to share common primary political units, relations among these units, power structures, stability of structures, primary decisional loci, and implementation styles. The three policies under consideration share such similarities. In each, the main societal actors at the inception of the policy regime were commercial interests (mining, logging, grazing); it was only later that environmental interests came to play a major role. Each policy regime was dominated by a western political clientele (not surprisingly, as these public lands are in the West). The policy regimes were overseen by two congressional committees (Agriculture and Interior, both of which have been categorized as clientele-oriented committees). And policy within each regime was implemented by an agency within either the Agriculture Department or the Interior Department, two departments also identified with their clientele. Based on these characteristics, we would expect these policy regimes to fall into the distributive category. Indeed, Lowi even mentions "natural-resources development" as an example of distributive policy. As natural resources policy began to attract the attention of a broader public, we might have expected each of these policy regimes

to move into the regulatory category. But, each policy regime has had a different pattern of politics (Lowi, "American Business, Public Policy," p. 688 specifically, pp. 677–715 generally). James Wilson also offers a policy typology of sorts, but it is a typology of agencies rather than policy, that is, it is geared to explaining how particular agencies function, rather than how particular policy regimes function (Wilson, *Bureaucracy*, pp. 72–89).

12. Clarke and McCool, *Staking Out the Terrain*; Clary, *Timber and the Forest Service*; Culhane, *Public Lands Politics*; Durant, *Administrative Presidency Revisited*; Wilkinson, *Crossing the Next Meridian*. An example of the type of analysis that is most useful is Hays, *Conservation and the Gospel of Efficiency*. This work is theoretical and comparative in its examination of natural resources policy over a thirty-year period.

13. By ideas, I mean thoughts, plans, schemes that are well formed, consistent, and coherent. As such, they fall between a developed ideology or public philosophy, which addresses the broad nature of society, and the material interests of specific societal groups. Ideas, then, can be thought of at the middle level, between an ideology (liberalism or Marxism) and the interests of a specific group (public-lands grazing fees should not be raised), though they are influenced by the former and influence the latter. For example, an idea regarding health care policy might be related to the larger philosophy of liberalism and might help to shape the interests of labor on a particular issue.

For more on ideas, see the Winter 1992 volume of *International Organization* on epistemic communities; Derthick and Quirk, *Politics of Deregulation*; Hall, "Conclusion"; Quirk, "In Defense of the Politics of Ideas"; Quirk, "Deregulation and the Politics of Ideas in Congress"; Weir, *Politics and Jobs*; Wood, "Rhetoric and Reality in the American Revolution."

14. Bennett and Sharpe, *Transnational Corporations Versus the State*, pp. 9–13, 43–47; Weir, *Politics and Jobs*, p. 21.

15. The state is defined to consist of those individuals who hold positions in government, including bureaucrats, executive officials, judges, and legislators. State capacity, synonymous with administrative capacity, is based on a number of factors: number of employees, size of budget, and professionalization of the agency. In addition, certain state structure characteristics will be factored into the estimation of state capacity, including centralization of policy authority, age of agency, and agency-congressional relations. State structure refers to a host of characteristics describing the arena of the state and the arena in which politics occurs. State structure incorporates separation of powers, the roles of committees in Congress, the powers of the judicial system, property rights, the design of the bureaucracy (including policy centralization and policy fragmentation), and the institutionalization of past policies. State structure is the terrain on which political battles are fought, both by agencies and interest groups. As in all battles, the terrain shapes strategies and outcomes. In the American context it is important to recognize the fragmented nature of the state. The definition of the state will not change from issue area to issue area, but the relevant actors and structure will change. Relatedly, state capacity will also vary from issue area to issue area. This variation is based on the history, nature, and structure of individual states.

Within the now vast literature on the state, among the most important works are Evans, Rueschemeyer, and Skocpol, *Bringing the State Back In*; March and Olsen, "New Institutionalism"; Nordlinger, *On the Autonomy of the Democratic State*; Skowronek, *Building a New American State*.

Chapter Two

1. Hurst, *Law and the Conditions of Freedom*, pp. 3–32.

2. Lowi, *End of Liberalism*, pp. 3–63.

3. House Committee on Public Lands and Smith as quoted in Peffer, *Closing of the Public Domain*, pp. 70, 94; Ballinger as quoted in Shanks, *This Land Is Yours*, p. 276; Peffer, *Closing of the Public Domain*, pp. 39–41.

4. Dana, *Forest and Range Policy*, pp. 213, 314–16.

5. Graf, *Wilderness Preservation and the Sagebrush Rebellions*, pp. 166–70; Knous, "Use of Public Lands," p. 211; Peffer, *Closing of the Public Domain*, pp. 279–93.

6. Aspinall as quoted in Gates, "Pressure Groups and Recent American Land Policies," p. 111.

7. The forefather of this movement may be economist Milton Friedman, who advocated the privatization of the national park system in the early 1960s (see Friedman, *Capitalism and Freedom*, p. 31). See Short, *Ronald Reagan and the Public Lands*, pp. 81–99.

8. Gates, *History of Public Land Law Development*, pp. 28–30; McConnell, "Conservation Movement," p. 473; Peffer, *Closing of the Public Domain*, p. 156.

9. Dana, *Forest and Range Policy*, pp. 213, 344, 232–36.

10. The Sagebrush Rebellion is discussed in more detail in Chapter 4.

11. Lowi, *End of Liberalism*, pp. 22, 69; McConnell, *Private Power and American Democracy*, p. 343. For a discussion of the theory that business, including the forest-products industry, often favored regulation as a way to reduce competition and increase order and profits, see Kolko, *Triumph of Conservatism*, generally and p. 111; Peffer, *Closing of the Public Domain*, p. 57.

12. See Gottlieb, *Wise Use Agenda*; Hagenstein, "Federal Lands Today," p. 97; Leshy, "Sharing Federal Multiple Use Lands," p. 270; Sax, "Claim for Retention of the Public Lands," p. 128.

13. Alston, *Individual and the Public Interest*, pp. 119–41; Behan, "Forestry and the End of Innocence," pp. 16–19, 38–49.

14. Fernow as quoted in Behan, "Myth of the Omnipotent Forester," p. 399; Behan, "Myth of the Omnipotent Forester," p. 399. For discussions of the technocratic nature of the early Forest Service, see Clary, *Timber and the Forest Service*, pp. 16, 25; Hays, *Conservation and the Gospel of Efficiency*, pp. 2, 3, 267, 271; Twight, *Organizational Values and Political Power*, pp. 15–24.

15. Pinchot, *Fight for Conservation*, pp. 60, 116; Pinchot as quoted in Twight, *Organizational Values and Political Power*, p. 25.

16. Pinchot and Graves as quoted in Dana, *Forest and Range Policy*, pp. 212, 208; Clepper, *Professional Forestry in the United States*, p. 66.

17. Pinchot, *Breaking New Ground*, p. 261.

18. Pinchot, *Fight for Conservation*, pp. 42–49. For more on Pinchot's utilitarianism, see Hays, *Conservation and the Gospel of Efficiency*, pp. 41–42; Twight, *Organizational Values and Political Power*, p. 7.

19. Pinchot as quoted in Nash, *Wilderness and the American Mind*, p. 161. For more on Pinchot's opposition to preservation, see Hays, *Conservation and the Gospel of Efficiency*, pp. 41–42; Twight, *Organizational Values and Political Power*, p. 7.

20. Roosevelt as quoted in Allin, *Politics of Wilderness Preservation*, p. 39; as quoted in Clary, *Timber and the Forest Service*, p. 21.

21. Clary, *Timber and the Forest Service*, p. 28; *A National Plan for American Forestry* as quoted in Clawson, *Federal Lands Revisited*, p. 130; Marshall as quoted in Alston, *Individual and the Public Interest*, p. 25.

22. McArdle as quoted in Clary, *Timber and the Forest Service*, p. 169; forestry professor quoted in Behan, "Myth of the Omnipotent Forester," p. 398.

23. Executive vice president of the SAF, H. R. Glascock, as quoted in Society of American Foresters, *Forestry for America's Future*, p. 11; Tuchmann, "Statement of E. Thomas Tuchmann," p. 2.

24. Clepper, *Professional Forestry in the United States*, p. 127 specifically, pp. 2–3, 123–34 generally; Kaufman, *Forest Ranger*; Twight, *Organizational Values and Political Power*, pp. 15–22. As Clepper writes, "The national forests created a need for foresters that led to the founding of schools of forestry, which in turn led to the formation of a profession of forestry" (*Professional Forestry in the United States*, p. 2). It should be noted that there has been much criticism of the dominance of technocratic utilitarianism within the Forest Service. On timber famine critiques, see Clary, *Timber and the Forest Service*; Olson, *Depletion Myth*. On interest-group capture, see McConnell, "Conservation Movement"; McConnell, *Private Power and American Democracy*; Reich, *Bureaucracy and the Forests*. And on the role of organizational values limiting the vision of foresters and the Forest Service, see Alston, *Individual and the Public Interest*; Behan, "Forestry and the End of Innocence"; Behan, "Myth of the Omnipotent Forester"; Schiff, *Fire and Water*; Schiff, "Innovation and Administrative Decision Making."

25. See Kessler et al., "New Perspectives for Sustainable Natural Resources Management," p. 224, for evidence of both the continued privilege of technocratic utilitarianism within the Forest Service and the rise of alternative ideas.

26. Nash, *Wilderness and the American Mind*, pp. 8–66.

27. Ibid., p. 67 specifically, pp. 67–83 generally.

28. Thoreau as quoted in ibid., p. 84; ibid., pp. 88, 89 specifically, pp. 84–95 generally. See also Oelschlaeger, *Idea of Wilderness*, pp. 133–71.

29. Nash, *Wilderness and the American Mind*, pp. 96–121.

30. Ibid., pp. 122–60. See also Fox, *American Conservation Movement*; Oelschlaeger, *Idea of Wilderness*, pp. 172–204; Runte, *National Parks*. For a history of the Sierra Club, see Cohen, *History of the Sierra Club*. Muir played a role in the establishment of Sequoia, Yosemite, Mount Rainier, Crater Lake, Glacier, and Mesa Verde National Parks, and Grand Canyon and Olympic National Monuments.

31. Muir as quoted in Nash, *Wilderness and the American Mind*, p. 161; ibid., pp. 161–81; Runte, *National Parks*. Ironically, in 1987, Interior Secretary Donald Hodel suggested draining the Hetch Hetchy reservoir to relieve pressure from the

overcrowded Yosemite Valley (see Shabecoff, "Historic Battle over a Yosemite Lake Is Back," p. A16).

32. Leopold as quoted in Nash, *Wilderness and the American Mind*, p. 186; ibid., pp. 182–99, 208–9. See also Leopold, *Sand County Almanac*; Meine, *Aldo Leopold*; Oelschlaeger, *Idea of Wilderness*, pp. 205–42.

33. As quoted in Fox, "We Want No Straddlers," p. 7; as quoted in Nash, *Wilderness and the American Mind*, p. 207; ibid., pp. 200–209. On Marshall and wilderness, see Marshall, "Problem of Wilderness"; Mitchell, "In Wildness Was the Preservation of a Smile"; Glover, *Wilderness Original*. For a history of the Wilderness Society, see Fox, "We Want No Straddlers."

34. Allin, *Politics of Wilderness Preservation*, pp. 102–42; Nash, *Wilderness and the American Mind*, pp. 209–26. The passage of the Wilderness Act will be discussed more fully in Chapters 3 and 4.

35. Zahniser as quoted in Nash, *Wilderness and the American Mind*, p. 233; ibid., pp. 227–37; Palmer, *Endangered Rivers and the Conservation Movement*.

36. The Wilderness Society as quoted in Nash, *Wilderness and the American Mind*, p. 298; Seiberling as quoted in ibid., p. 309; Kauffman as quoted in ibid., p. 308; ibid., pp. 272–315.

37. MacKaye as quoted in Nash, *Wilderness and the American Mind*, p. 245; Olson as quoted in ibid., p. 245; ibid., p. 264 specifically, pp. 238–71 generally.

38. Stegner as quoted in Nash, *Wilderness and the American Mind*, p. 262. For a discussion of the importance of wilderness in American literature, see Marx, *Machine in the Garden*.

39. See Kreiger, "What's Wrong with Plastic Trees?" In Chapter 6, I present a brief discussion of this new ecocentric and deep ecology thinking as it relates to preservationism.

40. As quoted in Baden and Lueck, "Property Rights Approach to Wilderness Management," pp. 47, 48, 50.

41. Other frameworks have been developed, but they have been descriptive rather than analytic. That is, they do not employ ideas as an explanatory variable, but rather identify categories within which various actors can be placed. There is some overlap between these frameworks and the ideas just presented, but for the purposes of explanation, I think that my approach is superior. See Alston, *Individual and the Public Interest*, pp. 32–33: economists, public foresters, preservationists; Culhane, *Public Lands Politics*, pp. 2–10, 20–21: utilitarianism, progressive conservationism, romantic preservationism, environmentalism; Fox, *American Conservation Movement*, pp. 107–9: amateurs (preservationists) and professionals (technocrats); Oelschlaeger, *Idea of Wilderness*, pp. 281–319: resourcism, preservationism, biocentrism and ecocentrism, deep ecology, ecofeminism; Petulla, *American Environmentalism*, pp. 24–39: biocentric, ecological, economic.

Chapter Three

1. Placer mining is done primarily in and along river and stream beds. Miners sought to recover gold or silver that had been eroded from larger deposits. Early California mining was almost exclusively placer mining. Quartz or lode mining is

based on removing minerals from hard rock. Usually, miners discovered a surface lode and then tunneled to follow the lode underground to recover minerals. After placer mines were played out, the more capital-intensive quartz mines became more important.

2. Mason as quoted in Leshy, *Mining Law*, p. 13.

3. Mason as quoted in Umbeck, *Theory of Property Rights*, pp. 69–70; Hershiser, "Influence of Nevada," pp. 132–35; Libecap, *Evolution of Private Mineral Rights*, pp. 190–91.

4. Paul, *Mining Frontiers of the Far West*, pp. 22–23; Umbeck, *Theory of Property Rights*, pp. 4–5, 73.

5. Umbeck, *Theory of Property Rights*, pp. 79–87.

6. Ellison, "Mineral Land Question in California," pp. 81–82; Paul, *Mining Frontiers of the Far West*, pp. 23–24; Shinn, "Land Laws of Mining Districts," pp. 555, 581; Shinn, *Mining Camps*, pp. 105–22, 232–58; Swenson, "Legal Aspects of Mineral Resources Exploitation," pp. 708–9; Umbeck, *Theory of Property Rights*, pp. 92–93, 95–96. See Leshy, *Mining Law*, pp. 379–80, for an outline of miners' rules; see Shinn, "Land Laws of Mining Districts," pp. 556–81, for a description of placer mining codes throughout California from 1848 to 1884, and pp. 601–5, 607–8 for examples of specific placer codes from California, Idaho, and Montana.

7. Ellison, "Mineral Land Question in California," pp. 72–73; Hershiser, "Influence of Nevada," pp. 135–36.

8. Ellison, "Mineral Land Question in California," pp. 74–75; Hershiser, "Influence of Nevada," pp. 138–41; Shinn, "Land Laws of Mining Districts," pp. 583–84; Swenson, "Legal Aspects of Mineral Resources Exploitation," pp. 712–13. The army had operated a lead-leasing program in the Midwest in the 1820s and 1830s that was judged unsuccessful. See Mayer and Riley, *Public Domain, Private Dominion*, pp. 20–37; Swenson, "Legal Aspects of Mineral Resources Exploitation," pp. 702–6.

9. Ellison, "Mineral Land Question in California," pp. 76–77.

10. Fillmore as quoted in ibid., p. 77, and in Swenson, "Legal Aspects of Mineral Resources Exploitation," p. 713.

11. Hurst, *Law and the Conditions of Freedom*, p. 7 specifically, pp. 3–32 generally.

12. Greever, *Bonanza West*, pp. 82–130; Libecap, *Evolution of Private Mineral Rights*; Libecap, "Economic Variables and the Development of the Law," pp. 338–47; Libecap, "Government Support of Private Claims"; Shinn, "Land Laws of Mining Districts," pp. 555, 585–88.

13. Ellison, "Mineral Land Question in California," pp. 82–83; Swenson, "Legal Aspects of Mineral Resources Exploitation," pp. 713–14.

14. Ellison, "Mineral Land Question in California," pp. 83–85; Hershiser, "Influence of Nevada," pp. 146–48, 151; Libecap, *Evolution of Private Mineral Rights*, pp. 193–95, 199–202; Libecap, "Economic Variables," p. 360; Mayer and Riley, *Public Domain, Private Dominion*, pp. 47–49; Swenson, "Legal Aspects of Mineral Resources Exploitation," pp. 714–16. Although these policies were debated in Congress, the executive branch played a role in developing and supporting proposals. It has even been reported that President Abraham Lincoln considered the question of mining policy the afternoon of his death (Leshy, *Mining Law*, p. 14).

15. Julian as quoted in Mayer and Riley, *Public Domain, Private Dominion*, p. 38; Ellison, "Mineral Land Question in California," p. 84; Swenson, "Legal Aspects of Mineral Resources Exploitation," pp. 714–15.

16. Conness as quoted in Shinn, "Land Laws of Mining Districts," p. 548; Stewart as quoted in Leshy, *Mining Law*, p. 15; Williams as quoted in Mayer and Riley, *Public Domain, Private Dominion*, p. 92; Ellison, "Mineral Land Question in California," p. 85; Hershiser, "Influence of Nevada," pp. 152–54, 162; Libecap, *Evolution of Private Mineral Rights*, pp. 196–98; Libecap, "Government Support of Private Claims," p. 372; Mayer and Riley, *Public Domain, Private Dominion*, pp. 49–52; Swenson, "Legal Aspects of Mineral Resources Exploitation," pp. 716–18. In addition, Senators Stewart and Conness succeeded in leading the passage of a law directing that federal court decisions on mining title not be affected by federal ownership of the land. According to Swenson, "This seems to have been the first Federal statute expressly recognizing, to a limited extent, an implied license to miners to go upon Federal land to extract minerals" (p. 717).

17. Ellison, "Mineral Land Question in California," pp. 85–87; Greever, *Bonanza West*, p. 107; Hershiser, "Influence of Nevada," pp. 162–63; Mayer and Riley, *Public Domain, Private Dominion*, pp. 52–53; Swenson, "Legal Aspects of Mineral Resources Exploitation," pp. 718–19.

18. 14 Stat. 251; 30 USC, secs. 21–54.

19. 16 Stat. 217; 17 Stat. 91, codified in 30 USC, secs. 21–54; Mayer and Riley, *Public Domain, Private Dominion*, p. 54; Swenson, "Legal Aspects of Mineral Resources Exploitation," pp. 721–23.

20. See Skowronek, *Building a New American State*.

21. In 1864 proposals were made in both the House and the Senate, and by the secretary of the interior, to create a mining bureau, a mining department, and a geological and mineralogical survey. None of these proposals was acted upon (Hershiser, "Influence of Nevada," p. 152).

22. For a general discussion of the development of property rights in the California and Nevada mineral rushes, see Libecap, *Evolution of Private Mineral Rights*; Libecap, "Economic Variables"; Libecap, "Government Support of Private Claims"; Umbeck, "California Gold Rush"; Umbeck, "Theory of Contract Choice and the California Gold Rush"; Umbeck, *Theory of Property Rights*; Umbeck, "Might Makes Rights."

23. Smith as quoted in Mayer and Riley, *Public Domain, Private Dominion*, pp. 183, 74; Dupree, *Science in the Federal Government*, pp. 195–214, 280–83; Mayer and Riley, *Public Domain, Private Dominion*, p. 74. A significant change in mining policy did develop regarding fossil fuels. The Mineral Leasing Act, passed in 1920, established a system for private companies to lease public lands containing coal, oil, oil shale, phosphate, potash, sodium, and sulfur and to pay the government a royalty for these resources. The Mineral Leasing Act represents a shift away from the economic liberalism idea of hard-rock mining. The leasing program, in which the government maintains some degree of control over mineral development, reflects a dominant theme of the time, a search for order. As will become apparent in the study of forestry, the leasing program shares much in common with the establishment of a national forest system administered by a body of experts guided by an

idea of what most furthered the public interest: technocratic utilitarianism. Despite this move to leasing for one type of mineral, however, the idea of economic liberalism remained firmly embedded in the hard-rock mining policy regime and the changes of 1920 did not affect it. See Hays, *Conservation and the Gospel of Efficiency*, pp. 82–90; Mayer and Riley, *Public Domain, Private Dominion*, pp. 114–40, 155–95; Swenson, "Legal Aspects of Mineral Resources Exploitation," pp. 724–45.

24. Leshy, *Mining Law*, pp. 230–31. This case focuses only on the Wilderness Act and mining. Chapter 4 contains a more comprehensive discussion of the passage of the Wilderness Act.

25. Roth, "National Forests and the Campaign for Wilderness Legislation," p. 121.

26. Senate, Committee on Interior and Insular Affairs, *Hearings on a Wilderness Preservation System* (1957), p. 7. Humphrey had introduced S. 4013, the first wilderness bill, in 1956 as a study bill. No hearings were held on the bill, though.

27. Ibid., p. 329.

28. A technocratic utilitarian approach to hard-rock minerals management never really developed in the United States. Rather, the mineral resource management professionals (e.g., geologists, mining engineers) aligned themselves with the economic liberals. These technical experts were employed in the private sector, and unlike their counterparts in forestry, they did not think that the private sector was failing to properly manage minerals. The free market was the proper tool for resource management.

29. Senate, Committee on Interior and Insular Affairs, *Hearing on a Wilderness Preservation System* (1958), pp. 5, 202.

30. Ibid., pp. 504, 225.

31. See, for example, ibid., pp. 609, 612, 789.

32. Senate, Committee on Interior and Insular Affairs, *Hearings on a Wilderness Preservation System* (1961).

33. Ibid., pp. 368–69.

34. Senate, "Report No. 635," p. 17.

35. House, Subcommittee on Public Lands, *Hearings on a Wilderness Preservation System* (1961), p. 583.

36. Ibid. (1962), pp. 1557, 1584–85, 1548.

37. Allin, *Politics of Wilderness Preservation*, pp. 127–29; House, "Report No. 2521," p. 118.

38. Senate, Committee on Interior and Insular Affairs, *Hearings on a Wilderness Preservation System* (1963).

39. Baker, "Conservation Congress of Anderson and Aspinall," pp. 108–10.

40. Baring as quoted in Allin, *Politics of Wilderness Preservation*, p. 131. Aside from S. 4 and H.R. 9162, the other major bill under consideration in the House was H.R. 9070, introduced by Saylor. The bill was similar to S. 4 except that it required affirmative congressional action before new wilderness areas could be designated. It included *no* exemption for mining activities and allowed for immediate designation for Forest Service primitive areas as wilderness. This bill was acceptable to preservationists, although they most favored S. 4. H.R. 9162 differed from H.R.

9070 on the mineral exemption and included no automatic designation of primitive areas as wilderness.

41. House, Subcommittee on Public Lands, *Hearings on a Wilderness Preservation System* (1964a); ibid. (1964b).

42. Baker, "Conservation Congress of Anderson and Aspinall," pp. 117–18.

43. Ibid., p. 118; Roth, "National Forests and the Campaign for Wilderness Legislation," p. 124; 78 Stat. 890; PL 88-577; 16 USC, secs. 1131–36.

44. Leshy, *Mining Law*, p. 232. Specifically, see 16 USC, secs. 1133 (d)(2) and (3).

45. See Leshy, *Mining Law*, pp. 232–33, for a discussion about the lack of mining in wilderness areas.

46. Ibid., pp. 233–35.

47. Neville as quoted in Frome, *Battle for the Wilderness*, p. 144; Allin, *Politics of Wilderness Preservation*, p. 200; *Izaak Walton League v. St. Clair*, 353 F. Supp. 698 (D. Minn. 1973).

48. Leshy, *Mining Law*, pp. 236–37, 238. As Leshy writes, "Without a dramatic shift in public opinion, or a crisis in strategic mineral supply, it is difficult to imagine such a change taking place [regarding mining in the wilderness]. Despite the compromise language in the Wilderness Act, as a practical matter wilderness was almost completely withdrawn from mining in 1964, and that status quo will be hard to dislodge" (p. 238).

49. Ibid., p. 235; Sumner, "Wilderness and the Mining Law," pp. 8–18.

50. Different definitions of strategic minerals were used throughout the policy debates. The typical factors weighed to determine if a mineral were strategic included its military importance, its economic importance, and, most important, United States import dependence (and hence, vulnerability). For example, in 1982, the United States imported more than 75 percent of thirteen minerals (in order of highest percentage imported): columbium, industrial diamonds, graphite, mica, strontium, manganese, bauxite and alumina, cobalt, tantalum, chromium, fluorspar, platinum group metals, and nickel; a 1983 Bureau of Mines report identified eight critical minerals based on their economic and military importance (chromium, cobalt, columbium, nickel, platinum, tantalum, titanium, and tungsten); and an Office of Technology Assessment report done in 1985 focused on import vulnerability for four minerals (chromium, cobalt, manganese, and platinum group metals). See Anderson, *Strategic Minerals*, p. 25; Office of Technology Assessment, *Strategic Materials*; Tilton and Landsberg, "Nonfuel Minerals," p. 57. On minerals price increases, see Eckes, *United States and the Global Struggle for Minerals*, p. 237; "Now the Squeeze on Metals," p. 50.

51. Leshy, *Mining Law*, pp. 183–228. For a historical discussion of the attitudes of the mining industry toward the environment, see Smith, *Mining America*.

52. Cameron, *At the Crossroads*, p. 215; "Dig It," p. 84; Hammer, "People and Business," p. 55; Mikesell, *Nonfuel Minerals*, p. 193; Office of Technology Assessment, *Management of Fuel and Nonfuel Minerals*, pp. 215–20, 337–38; Tilton, *Future of Nonfuel Minerals*, pp. 56–57.

53. Flawn, "Impact of Environmental Concerns on the Minerals Industry," p. 97.

54. Overton as quoted in Hammer, "People and Business," p. 55; Slappey,

"We're Headed for a Metals Crunch," pp. 21–24; Wade, "Raw Materials," pp. 185–86; "A Minerals Cartel Would Be Worse than the Energy Crisis," pp. 48–49; "Dig It," p. 84.

55. As discussed in the previous minerals policy case studies, in the hard-rock mining realm the technocratic utilitarian perspective of the public interest was virtually the same as the economic liberalism perspective. Hence, the geologists and mining interests formed a nearly permanent alliance on questions of public-lands minerals policy. It was no different in the strategic minerals case. Three different professional groups—the American Institute of Professional Geologists, the Mineral Exploration Coalition, and the Society of Economic Geologists—testified repeatedly in favor of increased access to the public lands for mining, including in wilderness areas. Some even advocated opening up national parks for mining. For example, the dean of the University of Nevada School of Mines, in his testimony concerning the proposed Great Basin National Park in Nevada, argued that all lands should be open to mineral exploration and development, even Yellowstone and Yosemite. Geologists within the government shared a similar perspective. As Mayer and Riley report, "The special role that their [USGS geologists] profession plays within the mining industry prevents them as regulators from challenging or even carefully monitoring mining operations on public lands. Long-standing historical and economic relationships make government geologists act as industry advocates." See Mayer and Riley, *Public Domain, Private Dominion*, p. 272; Shanks, *This Is Your Land*, p. 112.

56. See a series of articles on this issue in *National Parks and Conservation Magazine* (May 1975, December 1975, January 1976, November 1976) and the *New York Times* (October 6, 1975, pp. 1, 28, October 7, 1975, p. 36, May 9, 1976, sec. 4, p. 14).

57. Carter, "Minerals and Mining," pp. 809–11; Senate, Subcommittee on Science, Technology, and Space, *Hearings on Material Policy* (1977); House, Subcommittee on Science, Research and Technology, *Hearings on a National Policy for Materials* (1978); ibid. (1979).

58. Overton, "Mining Industry," p. 254; Shaine, "Alaska's Minerals," pp. 27–28.

59. House, Subcommittee on Mines and Mining, *Oversight Hearings on Nonfuel Minerals Policy Review*.

60. House, Subcommittee on Mines and Mining, *Report on U.S. Minerals Vulnerability*, pp. 68–70, 74–75. The claim of an antimineral bias disappeared with the Reagan administration. A BLM draft policy in 1987 recommended that the bureau "act as a minerals advocate in its coordination with other agencies and seek to reduce regulatory burdens and paperwork requirements affecting the minerals industry; . . . [and to recognize that] residual and/or unavoidable impacts from development are often to be considered acceptable given the benefits that result from mineral development." Although BLM director Robert Burford rejected the recommendations, the case for an antiminerals bias within Interior would seem to have evaporated. See "Agency Rejects Industry Role," p. A20. Preservationists challenged this perspective of state bias against mining, though, arguing that just the opposite was true. They cited a 1974 NASA study of minerals management: "We [NASA] found strong biases almost everywhere in favor of production and the interests of the mineral industry—often at the expense of other valid objectives,

such as protection of the environment, seeking maximum ultimate recovery, and getting fair market value." See Mayer and Riley, *Public Domain, Private Dominion*, p. 273.

61. House, Subcommittee on Mines and Mining, *Report on U.S. Minerals Vulnerability*, pp. 87–88.

62. Senate, Subcommittee on Energy Resources and Materials Production, *Hearings on Materials Policy, Research, and Development Act*; Senate, Subcommittee on Science, Technology, and Space, *Hearings on a National Materials Policy*, pp. 37, 93.

63. 94 Stat. 2305; PL 96-479; 30 USC, secs. 1601–5.

64. McClure as quoted in Hershey, "New Senate Energy Chief," p. D6; Anderson, *Structure and Dynamics of U.S. Government Policymaking*, p. 50. See also Goldwater, "U.S. Dependency on Foreign Sources for Critical Material," for a similar viewpoint from another senior Republican senator. This combination of public and governmental interest at the time led Anderson to comment that "during no other period, in peacetime at least, had there been a greater interest, both public and governmental, in strategic minerals as at the time Reagan assumed the presidency" (p. 50).

65. Watt as quoted in "Watt Vows Shift on Key Minerals," p. B10; "Federal Land Urged as Strategic Mineral Source," pp. 50–51; Shabecoff, "Watt to Seek Ban," p. A1. See also Watt's testimony concerning strategic minerals and the public lands in Senate, Subcommittee on Energy and Mineral Resources, *Hearings on Strategic Minerals and Materials Policy*.

66. "Interior Dept. Seeks Mining Permits in Parks," p. A10; Shabecoff, "Drilling and Mining Planned," p. A33.

67. Senate, Subcommittee on Energy and Mineral Resources, *Hearings on Strategic Minerals and Materials Policy*, pp. 64, 66. See also the exchange in the *New York Times*: "Editorial: The Minerals Problem Is Not a Crisis," p. A22; Overton, "Letter," p. A14.

68. H.R. 3364.

69. House, Subcommittee on Mines and Mining, *Hearings on National Minerals Security Act*. See also "Editorial: The Minerals Problem Is Not a Crisis," p. A22; Shanks, *This Is Your Land*, pp. 117–18.

70. It should be recalled that the Wilderness Act mandated USGS–Bureau of Mines surveys on the mineral potential of wilderness lands. A 1982 USGS report on sixty wilderness and Forest Service primitive areas indicated that 60 percent of the areas studied had little or no economic mineral potential, 22 percent had moderate potential, and 18 percent had relatively high potential. By 1984, the USGS and Bureau of Mines had assessed the mineral potential of approximately eight hundred areas on forty-five million acres of wilderness and wilderness study areas, primarily on national forest lands. The report contained no overall summary of the mineral potential of these lands, nor any assessment of the potential for strategic minerals on these lands. A number of important finds in wilderness and wilderness study areas were reported occasionally, with the most promising finds being reported in Alaska, where geologists thought that a mineral belt might exist north of the Brooks Range. According to one analyst, the wilderness issue took on such importance because "the mining industry fought what it saw as a threat to

close mining altogether out of public lands. . . . Many in the industry seemed to assume that the only place one could find strategic minerals would be in areas designated for wilderness." In seeming support of this thinking, as of 1982 there were an estimated fifty thousand hard-rock mineral claims and one thousand oil and gas applications in wilderness areas, even though there had been no development of minerals allowed in these areas. See Cameron, *At the Crossroads*, pp. 212–19; Curlin, "Political Dimensions of Strategic Minerals," p. 141; Marsh, Kropschot, and Dickinson, *Wilderness Potential*; "Metals Found on Oil Reserve," p. D6; Mikesell, *Nonfuel Minerals*, pp. 193–98; "Mining Hopes in New Mexico," p. 38; Shabecoff, "Debate over Wilderness Area Leasing Intensifies," p. D6; "U.S. Reports Discovery of Minerals in Alaska," p. 40.

71. Mayer and Riley, *Public Domain, Private Dominion*, pp. 255–56, 260; Shabecoff, "Debate over Wilderness Area Leasing Intensifies," p. D6; Shabecoff, "Watt to Seek Ban," p. A1.

72. Shabecoff, "Bill Would Bar Wilds Drilling," p. A22.

73. "Controversy over Wilderness Area Minerals Policy"; Curlin, "Political Dimensions of Strategic Minerals," pp. 135–47; Mayer and Riley, *Public Domain, Private Dominion*, pp. 260–62.

74. The report is reprinted in Senate, Subcommittee on Energy and Mineral Resources, *Hearing on the President's National Materials and Minerals Program and Report to Congress*, pp. 2–24 generally, pp. 3, 7, 8 specifically. See also Shabecoff, "Reagan Tells of Plan," p. A1.

75. Senate, Subcommittee on Energy and Mineral Resources, *Hearing on the President's National Materials and Minerals Program and Report to Congress*, p. 34.

76. Ibid., pp. 103–4, 136, 189.

77. Shabecoff, "Reagan Tells of Plan," p. A1.

78. "The Strategic Minerals Fallacy," pp. 40–41.

79. House, Subcommittees on Transportation, Aviation, and Materials and Science, Research and Technology, *Oversight Hearings on PL 96-479 and Consideration of HR 4281, Critical Materials Act of 1981*, pp. 76, 78, 81.

80. Raloff, "Watt Yields on Wilderness Leasing," p. 21.

81. 98 Stat. 1248; Title II of PL 98-373; 30 USC, secs. 1801–11. President Reagan appointed Interior Secretary Donald Hodel to chair the council (the other members of the council were from the defense and energy departments). The council prepared a report on the critical materials issue as part of its responsibility: Executive Office of the President, National Critical Materials Council, "A Critical Materials Report." See also Anderson, *Structure and Dynamics of U.S. Government Policymaking*, pp. 34–36, 93–105.

82. Shabecoff, "U.S. Agency Drops Its Rule on Mines," p. A14; Shabecoff, "U.S. Cuts Off Protection of Millions of Acres," p. A1.

83. House, Subcommittee on Transportation, Aviation, and Materials, *Hearings on the National Critical Materials Act of 1984*. See also House, Subcommittees on Transportation, Aviation, and Materials and Science, Research and Technology, *Oversight Hearings on PL 96-479 and Consideration of HR 4281, Critical Materials Act of 1981*.

84. Anderson, *Structure and Dynamics of U.S. Government Policymaking*, p. 66; Cook, "Crisis That Didn't Happen," pp. 91–94.

85. Anderson, *The Structure and Dynamics of U.S. Government Policymaking*; Netschert, "Better Management of Nonfuel Minerals on Federal Land," pp. 193–95; "Problems Found in Mineral Rights," p. 33. Fragmentation also existed in Congress, where four House committees and two Senate committees held hearings on the strategic minerals issue during this period, and in state-federal relations due to differing state regulatory regimes, which further fragment state coherence and authority. See Leshy, *Mining Law*, pp. 212–20.

86. Leshy, *Mining Law*, pp. 195–99, 343.

87. Julian as quoted in ibid., p. 361.

Chapter Four

1. Dana and Fairfax, *Forest and Range Policy*, pp. 41–43, 50–51; Meyer, "Forests and Forestry," p. 43. In many ways, the publication of George Perkins Marsh's *Man and Nature* in 1864 marks the beginning of national concern with conservation issues. In his book, Marsh analyzed the importance of natural resources in supporting civilization and reported on the decline of numerous civilizations that did not protect or maintain their resources.

2. Van Hise, *Conservation of Natural Resources*, p. 214.

3. 26 Stat. 1095; 16 USC, sec. 471 (repealed); Dana and Fairfax, *Forest and Range Policy*, pp. 55–58. There is some debate in accounts of the passage of the bill as to how it fared on the floor. Gifford Pinchot and Robert Yard, in their accounts, claim that members of Congress, especially the opponents of forest reserves, did not expect any forestry measures in a general lands act, especially since adding new programs to bills in conference committee violated a congressional rule. In the typical rush at the end of a congressional session, the bill and rider passed without opposition. Samuel Dana and Harold Steen, alternatively, report that members of Congress knew that the bill contained the forest-reserve provision, but wanted to pass the overall bill and claimed that they could pass a bill or joint resolution overriding any presidential reserve. The evidence tends to support Dana and Steen. See Dana, *Forest and Range Policy*, p. 101; Pinchot, *Breaking New Ground*, p. 85; Steen, *U.S. Forest Service*, pp. 26–27; Yard, *Our Federal Lands*, pp. 109–10.

4. Pinchot, *Breaking New Ground*, pp. 108–10; Yard, *Our Federal Lands*, p. 111.

5. 30 Stat. 11, 34; 16 USC, sec. 475; Dana and Fairfax, *Forest and Range Policy*, pp. 58–62, 91–92; Pinchot, *Breaking New Ground*, pp. 113–19; Van Hise, *Conservation of Natural Resources*, p. 215; Yard, *Our Federal Lands*, pp. 111–12. The forest reserves were renamed national forests in 1907.

6. There is some controversy over the role of conservation within the larger progressive movement. Samuel Hays argues that conservation differed from the movement in that it was not antimonopoly and that it favored expert control, not democratic control. J. Leonard Bates argues that conservation was part of the progressive movement and the general efforts to reform government, reduce corruption, increase democratic control, and break monopoly capitalism. I think

that Hays makes the stronger case. See Bates, "Fulfilling American Democracy"; Hays, *Conservation and the Gospel of Efficiency*, pp. 2–3, 263–65.

7. Wiebe, *Search for Order*, pp. 160, 193 specifically, pp. 145–95 generally. Hays concurs in this interpretation of the period as one involving the search for increased order (*Conservation and the Gospel of Efficiency*, pp. 261–76).

8. Fox, *American Conservation Movement*, pp. 121, 129.

9. Dana and Fairfax, *Forest and Range Policy*, pp. 80–81; Hays, *Conservation and the Gospel of Efficiency*, pp. 38–45; Pinchot, *Breaking New Ground*, pp. 198–201, 254–56; Steen, *U.S. Forest Service*, pp. 60–61, 71–72.

10. Pinchot as quoted in Ponder, "Gifford Pinchot," p. 26; Ponder, "Gifford Pinchot," pp. 26, 35; Ponder, "Federal News Management in the Progressive Era," p. 47; Peffer, *Closing of the Public Domain*, pp. 66–69.

11. Fox, *American Conservation Movement*, p. 129; Ponder, "Gifford Pinchot," pp. 26–35; Ponder, "Federal News Management in the Progressive Era," pp. 42–48.

12. The introduction to the minority report reads: "We are of the opinion that this very radical change in policy is impracticable, would be unwise, and exceedingly expensive; that it would lead to a very great friction between the Agriculture and the Interior departments; would necessitate the creation of many new offices; would be detrimental to the interests of the Government and of the people as a whole, as well as those living in the vicinity of the reserves." See House, "Report No. 968," p. 1.

13. Hays, *Conservation and the Gospel of Efficiency*, p. 41.

14. 33 Stat. 628; 16 USC, sec. 472; Hays, *Conservation and the Gospel of Efficiency*, p. 44.

15. Dana and Fairfax, *Forest and Range Policy*, p. 81; Ise, *United States Forest Policy*, p. 158; Peffer, *Closing of the Public Domain*, pp. 92–93.

16. Hays, *Conservation and the Gospel of Efficiency*, pp. 44–45; Pinchot, *Breaking New Ground*, pp. 261–62.

17. Dana and Fairfax, *Forest and Range Policy*, pp. 79, 83; Dupree, *Science in the Federal Government*, pp. 245, 249; Harmon, "What Should Foresters Wear?"; McConnell, *Private Power and American Democracy*, p. 44; Pinchot, *Breaking New Ground*, pp. 281–305; Pinkett, "Forest Service, Trail Blazer in Recordkeeping Methods"; U.S. Department of Agriculture, *Annual Reports of the Department of Agriculture: 1900*, (p. 103), *1902* (p. 386), *1904* (p. 317), *1906* (p. 642), *1907* (p. 62), *1908* (p. 775). Illustrative of the interwoven nature of the forestry profession, the deans of the first three forestry schools—Cornell, Yale, Michigan—all served in high positions at the Division or Bureau of Forestry.

18. For a more complete discussion of Forest Service autonomy, see Klyza, "Window of Autonomy." On the lack of state capacity at the time generally, see Skowronek, *Building a New American State*. The competition between the Departments of Agriculture and the Interior has continued to this day.

19. Peffer, *Closing of the Public Domain*, pp. 93–96; Pinchot, *Fight for Conservation*, p. 116; Ponder, "Gifford Pinchot," pp. 34–35; Roosevelt, *Autobiography*, pp. 439–40.

20. By relative autonomy, I mean that the Forest Service's actions were con-

strained by societal and legal factors, but that it still had significant discretion within these constraints. Prior to this period, societal and legal factors were weak constraints on the agency's actions.

21. A comprehensive intellectual history of the idea of wilderness in American culture is Nash, *Wilderness and the American Mind*. Additional important works on the role of wilderness in American culture are Marx, *Machine in the Garden*, on the importance of wilderness in American literature; Oelschlaeger, *Idea of Wilderness*, on the overall development of the idea of wilderness with special attention to major American figures; Smith, *Virgin Land*, on the sociological influences of a wilderness west on American culture. For a comprehensive history of the development of the national parks and the National Park Service, see Runte, *National Parks*.

22. Dana and Fairfax, *Forest and Range Policy*, pp. 132–34; Frome, *Forest Service*, pp. 95–97; Leopold, "Wilderness and Its Place in Forest Recreational Policy."

23. Dana, *Forest and Range Policy*, pp. 269–71; Dana and Fairfax, *Forest and Range Policy*, pp. 155–57; Steen, *U.S. Forest Service*, pp. 156–62, 209–13. Also see Twight, *Organizational Values and Political Power*, for a case study of the Forest Service–NPS conflict.

24. Nash, *Wilderness and the American Mind*, p. 207.

25. Dana and Fairfax, *Forest and Range Policy*, pp. 157–58.

26. Ibid., p. 198; Hession, "Legislative History of the Wilderness Act," pp. 31–32.

27. Nash, *Wilderness and the American Mind*, pp. 220–21.

28. Ibid., p. 219, and more generally, pp. 209–19.

29. Allin, *Politics of Wilderness Preservation*, pp. 104–6; Sundquist, *Politics and Policy*, pp. 337–38.

30. Allin, *Politics of Wilderness Preservation*, pp. 105–7; Nash, *Wilderness and the American Mind*, pp. 221–22.

31. Nash, *Wilderness and the American Mind*, p. 222. According to Nash, "Congress lavished more time and effort on the wilderness bill than on any other measure in American conservation history" (p. 222).

32. Allin, *Politics of Wilderness Preservation*, p. 108.

33. "Editorial: Forever Wild," p. ii; Senate, Committee on Interior and Insular Affairs, *Hearings on a Wilderness Preservation System* (1957), p. 354.

34. Allin, *Politics of Wilderness Preservation*, pp. 108–16; Nash, *Wilderness and the American Mind*, pp. 222–23; Sundquist, *Politics and Policy*, p. 338. The specific issue of mining and wilderness is discussed more thoroughly in Chapter 3.

35. Senate, Committee on Interior and Insular Affairs, *Hearings on a Wilderness Preservation System* (1957), p. 152; House, Subcommittee on Public Lands, *Hearings on a Wilderness Preservation System* (1962), pp. 1578–79.

36. House, Subcommittee on Public Lands, *Hearings on a Wilderness Preservation System* (1962), p. 1694.

37. Senate, Committee on Interior and Insular Affairs, *Hearings on a Wilderness Preservation System* (1957), pp. 9, 11.

38. Ibid., pp. 107–11, 379–80.

39. Ibid., pp. 281–89.

40. Allin, *Politics of Wilderness Preservation*, p. 123; Mercure and Ross, "Wilderness Act," pp. 54–55.

41. Allin, *Politics of Wilderness Preservation*, pp. 116–18; Mercure and Ross, "Wilderness Act," pp. 53–54.

42. Allin, *Politics of Wilderness Preservation*, pp. 118–19; Nash, *Wilderness and the American Mind*, p. 225.

43. Hession, "Legislative History of the Wilderness Act," pp. 54–70.

44. Allin, *Politics of Wilderness Preservation*, pp. 121–23.

45. Ibid., pp. 123–25; McCloskey, "Wilderness Act of 1964," p. 299.

46. Hession, "Legislative History of the Wilderness Act," pp. 128–30.

47. Allin, *Politics of Wilderness Preservation*, pp. 125–29; Mercure and Ross, "Wilderness Act," pp. 57–58; Sundquist, *Politics and Policy*, pp. 359–60. This was only the sixth time in history that a speaker had denied such a request of a committee chair.

48. Allin, *Politics of Wilderness Preservation*, pp. 129–31.

49. Hession, "Legislative History of the Wilderness Act," pp. 136–46.

50. Allin, *Politics of Wilderness Preservation*, pp. 131–34.

51. 78 Stat. 890; PL 88-577; 16 USC, secs. 1131–36; Allin, *Politics of Wilderness Preservation*, p. 135; Mercure and Ross, "Wilderness Act," p. 59; Sundquist, *Politics and Policy*, p. 361.

52. On clear cutting, see Clary, *Timber and the Forest Service*, pp. 180–94.

53. For a discussion of these transfer movements, see Graf, *Wilderness Preservation and the Sagebrush Rebellions*.

54. Clayton, "Sagebrush Rebellion"; Graf, *Wilderness Preservation and the Sagebrush Rebellions*, pp. 225–32; Leshy, "Unraveling the Sagebrush Rebellion"; Mollison and Eddy, "Sagebrush Rebellion"; Short, *Ronald Reagan and the Public Lands*, pp. 10–39.

55. Simmons and Baden, "Theory of the NRE"; Stroup and Baden, *Natural Resources*, p. 9. Not all economists favored privatization. For opposing views, see Bromley, "Public and Private Interests in the Federal Lands"; Runge, "Economist's Critique of Privatization"; Runge, "Fallacy of Privatization"; Runge, "Rejoinder."

56. Stroup and Baden, *Natural Resources*, pp. 111, 118; Dowdle, "Case for Privatizing Government Owned Timberlands," p. 83. On forestry, see Deacon and Johnson, *Forestlands*; on wilderness, see Baden and Lueck, "Property Rights Approach to Wilderness Management," and Stroup and Baden, "Endowment Areas"; on grazing, see Libecap, *Locking Up the Range*.

57. Hanke, "Privatization Debate," pp. 656–57, 662. See also the following works by Hanke: "Grazing for Dollars"; "Privatize Those Lands"; "On Privatizing the Public Domain"; "Wise Use of Federal Land," p. 31.

58. Lewis, "Reagan Administration's Federal Land Sales Program."

59. House, Subcommittee on Public Lands and National Parks, *Oversight Hearings on Public Land Sales and Transfers*, pp. 403–6.

60. Ibid. For a discussion of technical difficulties that the AMP would face, see General Accounting Office, *Numerous Issues Involved in Large-Scale Disposals and Sales of Federal Real Property*.

61. Lewis, "Reagan Administration's Federal Land Sales Program."

62. Senate, Committee on Energy and Natural Resources, *Hearings on Federal Property Management and Disposal*, pp. 121, 126, 139. See also House, Subcommittee on Public Lands and National Parks, *Oversight Hearings on Administration's Asset Management Program*, pp. 183, 185.

63. House, Subcommittee on Public Lands and National Parks, *Oversight Hearings on Public Land Sales and Transfers*, pp. 92–93; House, Subcommittee on Public Lands and National Parks, *Oversight Hearings on Administration's Asset Management Program*, p. 8.

64. Senate, Committee on Energy and Natural Resources, *Hearings on Federal Property Management and Disposal*, pp. 29, 30; House, Subcommittee on Public Lands and National Parks, *Oversight Hearings on Public Land Sales and Transfers*, p. 124.

65. As quoted in Stoler, "Land Sale of the Century," p. 22; House, Subcommittee on Public Lands and National Parks, *Oversight Hearings on Administration's Asset Management Program*, p. 27. As legal scholar Joseph Sax argues, "Current users, many of whom have used public lands at modest prices and benefited from public subsidies, have much to lose—particularly in a time of high interest rates—if they must bid against large investors and foreign capital for the continued use of lands now federally owned. For this reason, they are understandably reluctant about market-price sales into private ownership. . . . In short, it is quite possible that privatization would make current user constituencies of all stripes worse off, to the advantage of other remote interests, such as foreign investors and high bracket taxpayers" ("Why We Will Not [Should Not] Sell the Public Lands," pp. 314, 315).

66. Senate, Committee on Energy and Natural Resources, *Hearings on Federal Property Management and Disposal*.

67. Ibid., p. 96.

68. House, Subcommittee on Public Lands and National Parks, *Oversight Hearings on Public Land Sales and Transfers*.

69. Struck, "Block Wants to Sell Forest Land," p. A21; "Agency Seeks to Sell 10% of National Forests," p. 36. The sale of one of the Land Utilization parcels, the Hector Land Use Area in upstate New York, became so controversial that Congress reclassified the area as part of the Green Mountain National Forest in Vermont so that congressional authority would be needed to sell it off. See "Battle in the Finger Lakes," p. 50.

70. House, Subcommittee on Public Lands and National Parks, *Oversight Hearings on Public Land Sales and Transfers*, p. 138.

71. "Land Sale Goal Is Scaled Back," p. A21.

72. "Agency Weighing Sale of 3.2% of Forest System," p. 16; "6 Million Acres of U.S. Forest Eyed for Sale," p. A8.

73. As quoted in Stoler, "Land Sale of the Century," p. 22; Schmidt, "Skeptics in West Hear Case," p. 12.

74. Shabecoff, "Watt Removes Agency's Land," p. 1; "No Large Land Sales Planned, Watt Says," p. A18; "Surplus Western Lands May Still Be for Sale," p. A17.

75. Struck, "Sale of Surplus Lands a Bust," p. A13.

76. PL 98-146; 97 Stat. 953.

77. House, Subcommittee on Public Lands and National Parks, *Oversight Hearings on Public Land Management Policy*, p. 63.

Chapter Five

1. Dennen, "Cattlemen's Associations and Property Rights"; Foss, *Politics and Grass*, pp. 30–35; Hays, *Conservation and the Gospel of Efficiency*, pp. 50–52; Peffer, *Closing of the Public Domain*, pp. 22–26; Scott, "Range Cattle Industry."

2. Dana, *Forest and Range Policy*, pp. 115–16, 146–47; Hays, *Conservation and the Gospel of Efficiency*, pp. 44, 55–60, 253–54; Peffer, *Closing of the Public Domain*, pp. 72–77. This chapter will not be focusing upon Forest Service grazing policy but upon grazing policy on former public-domain lands now managed by the Bureau of Land Management. Despite the importance of grazing in national forests, it is not the central concern of the Forest Service nor the central reason behind the creation of the national forests. For more on grazing in the national forests, see Roberts, *Hoof Prints on Forest Ranges*; Rowley, *U.S. Forest Service Grazing and Rangelands*; Voigt, *Public Grazing Lands*, pp. 43–239.

3. Dana, *Forest and Range Policy*, p. 156; Foss, *Politics and Grass*, p. 42; Hays, *Conservation and the Gospel of Efficiency*, pp. 60–65; Peffer, *Closing of the Public Domain*, pp. 27–30, 72–98.

4. Theodore Lowi and Grant McConnell discuss the rise of interest-group liberalism in the 1930s, which was seen at work in agriculture perhaps above all other sectors. Lowi writes that agriculture "has been a system of self-government in which each leading farm interest controls a segment of agriculture through a delegation of national sovereignty" since this period. Although Lowi does not mention grazing in his book, it seems to illustrate his thesis ideally. McConnell uses the example of grazing policy and the Taylor Grazing Act to illustrate private expropriation of public authority. See Lowi, *End of Liberalism*, p. 69; McConnell, *Private Power and American Democracy*, pp. 200–211, on the Taylor Grazing Act.

5. Foss, *Politics and Grass*, p. 47; Peffer, *Closing of the Public Domain*, pp. 171–81, 190–93, 198–99; Voigt, *Public Grazing Lands*, pp. 246–48.

6. Foss, *Politics and Grass*, pp. 50–51; Peffer, *Closing of the Public Domain*, pp. 215–16; Voigt, *Public Grazing Lands*, pp. 249–50.

7. Stout, "Cattlemen, Conservationists, and the Taylor Grazing Act."

8. Peffer, *Closing of the Public Domain*, pp. 216–20; Stout, "Cattlemen, Conservationists, and the Taylor Grazing Act," pp. 321–28.

9. Ickes as quoted in Foss, *Politics and Grass*, p. 173; Dana and Fairfax, *Forest and Range Policy*, pp. 160–61; Foss, *Politics and Grass*, pp. 51–53, 57–58; Gates, *History of Public Land Law Development*, p. 612; Peffer, *Closing of the Public Domain*, pp. 216–20.

10. Rowley, *U.S. Forest Service Grazing and Rangelands*, p. 152.

11. 48 Stat. 1269; 43 USC, sec. 315; Calef, *Private Grazing and Public Lands*, pp. 52–57; Dana, *Forest and Range Policy*, pp. 259–62; Foss, *Politics and Grass*, pp. 59–60; Gates, *History of Public Land Law Development*, p. 611; Peffer, *Closing of the Public Domain*, pp. 221–24; Voigt, *Public Grazing Lands*, pp. 250–52. The act designated that the lands be classified as to their potential use, with agricultural lands to remain open for homesteading. President Roosevelt, via executive orders in 1934

and 1935, withdrew all remaining public lands for such classification, an action that essentially closed the public domain.

12. Carpenter as quoted in Hendricks, "Farrington R. Carpenter," p. 30; Foss, *Politics and Grass*, pp. 78–84; Gates, *History of Public Land Law Development*, pp. 614–15; Voigt, *Public Grazing Lands*, pp. 253–62.

13. Carpenter as quoted in Foss, *Politics and Grass*, p. 119; as quoted in ibid., pp. 127, 135–36, and more generally, pp. 127–32.

14. The Grazing Service was merged with the GLO to create the BLM, where it became the Division of Range Management, by Presidential Reorganization Plan Number 3. Dana and Fairfax, *Forest and Range Policy*, pp. 186–88; Foss, *Politics and Grass*, p. 86 particularly, pp. 127–36, 202 generally. See also McConnell, *Private Power and American Democracy*, p. 209. As Foss elaborates on the advisory boards, "The boards appear to be the dominant rule-making (policy-formulating) body in the federal grazing service. The boards are also involved in managerial details and, in fact, no part of the administration appears to be barred from their surveillance. The boards determine how range improvement funds are to be expended and their decisions on individual grazing permits are rarely overruled. The boards are influential in the selection of administrative personnel. The relationships between the boards and the two livestock producers organizations are so close that leadership in the associations and the boards is frequently embodied in the same persons" (*Politics and Grass*, pp. 135–36).

15. Calef, *Private Grazing and Public Lands*, pp. 59, 262; Dana and Fairfax, *Forest and Range Policy*, p. 163; Foss, *Politics and Grass*, pp. 84, 89, 92–97, 131, 200–202; Hendricks, "Farrington R. Carpenter"; Klemme, *Home Rule on the Range*, pp. 177–85, 190–93; McConnell, *Private Power and American Democracy*, p. 207; Peffer, *Closing of the Public Domain*, pp. 227–31; Penny and Clawson, "Administration of Grazing Districts."

16. Chapman, "Reorganization and the Forest Service"; Dana and Fairfax, *Forest and Range Policy*, p. 164; Gates, *History of Public Land Law Development*, pp. 615–17; Peffer, *Closing of the Public Domain*, pp. 232–42, 330–31; Polenberg, "Conservation and Reorganization." Indeed, the case of the Division of Grazing further supports Terry Moe's thesis on the politics of bureaucratic structure: the bureaucracy often is designed to function poorly (Moe, "Politics of Bureaucratic Structure").

17. Dana and Fairfax, *Forest and Range Policy*, pp. 286–87; Gates, *History of Public Land Law Development*, pp. 617–22; Peffer, *Closing of the Public Domain*, pp. 247–48.

18. Dana and Fairfax, *Forest and Range Policy*, p. 338; Senzel, "Genesis of a Law, Part 1," p. 32. For general discussions of grazing law during this period, see Kingery, "Public Grazing Lands"; Coggins, Evans, and Lindeberg-Johnson, "Law of Public Rangeland Management I"; Coggins and Lindeberg-Johnson, "Law of Public Rangeland Management II"; Coggins, "Law of Public Rangeland Management III"; Coggins, "Law of Public Rangeland Management IV"; Coggins, "Law of Public Rangeland Management V."

19. Dana and Fairfax, *Forest and Range Policy*, p. 231.

20. Pearl as quoted in "More Commercial Use of Public Land Urged," p. 5; Aspinall as quoted in Hill, "Revised Policy for U.S. Lands Asked in Study," p. 1.

For more general overviews of the PLLRC report, see Dana and Fairfax, *Forest and Range Policy*, pp. 231–35; "Symposium on the Public Land Law Review Commission Revisited"; Gates, "Pressure Groups and Recent American Land Policies"; "Symposium on the Public Land Law Review Commission." The report also reflected the original variant of economic liberalism, suggesting that much of the public lands should be transferred to the private sector to further the public good.

21. Berry, "Analysis," p. 19 (Sierra Club quotes); McCloskey, "Analysis," p. 21; National Wildlife Federation, quoted in "More Commercial Use of Public Land Urged," p. 5.

22. "News: PLLRC," p. 3; House, Subcommittee on the Environment, *Hearings on the Public Land Policy Act of 1971*; House, Committee on Interior and Insular Affairs, *Report to Accompany National Land Policy, Planning and Management Act of 1972*; Senate, Committee on Interior and Insular Affairs, "Public Land Policy: Activities in the 92nd Congress." Senator Gordon Allott (R, Colo.) introduced a similar bill in the Senate.

23. As quoted in "News: PLLRC," p. 3; Senate, Committee on Interior and Insular Affairs, *Hearings on Legislation to Revise the Public Land Laws*; Senate, Committee on Interior and Insular Affairs, "Public Land Policy: Activities in the 92nd Congress." The coalition consisted of the following groups: American Forestry Association, Defenders of Wildlife, Federation of Western Outdoor Clubs, Friends of the Earth, Izaak Walton League of America, National Association of State Foresters, National Audubon Society, National Recreation and Parks Association, National Wildlife Federation, North American Wildlife Foundation, Sierra Club, Society of Range Management, Sports Fishing Institute, Wilderness Society, Wildlife Management Institute, and Wildlife Society. Representative John Saylor (R, Pa.) sponsored a similar bill in the House.

24. Senzel, "Genesis of a Law, Part 1," p. 62; Senate, Committee on Interior and Insular Affairs, "Public Land Policy: Activities in the 92nd Congress."

25. Gates, "Pressure Groups and Recent American Land Policies," pp. 117–18; Senzel, "Genesis of a Law, Part 1," pp. 63–64.

26. "Editorial: Scuttling Land Reform," p. 36; Gates, "Pressure Groups and Recent American Land Policies," pp. 117–18; Lavender, "Decision for Permanence," p. 10; "NPCA at Work: BLM," pp. 28–29.

27. Senzel, "Genesis of a Law, Part 2," p. 34; Schwartz, "Capsule Examination."

28. House, Subcommittee on Public Lands, *Hearings on Public Land Policy and Management Act of 1975*. Seiberling had taken over the role of Representative Saylor, who died in October 1973.

29. Ibid. An AUM is the amount of forage needed for one cow and a calf or five sheep for one month.

30. Ibid., p. 32.

31. Ibid., p. 474. The groups offering a joint statement were: American Forestry Association, Boone and Crockett Club, Citizens Committee on Natural Resources, International Association of Game Fish and Conservation Commissioners, Izaak Walton League of America, National Audubon Society, National Rifle Association, National Wildlife Federation, Sport Fishing Institute, Wildlife Management Institute, and Wildlife Society.

32. House, Subcommittee on Public Lands, *Hearings on Public Land Policy and Management Act of 1975*, pp. 64, 66.

33. Alderson, "Capitol Watch," p. 41; Evans, "Washington Report," pp. 22–23.

34. "Public Land Management," pp. 186–87; Senzel, "Genesis of a Law, Part 2," pp. 36–37; "News: BLM Organic Act Passes House," p. 31.

35. Schwartz, "Capsule Examination," pp. 297–300; Senzel, "Genesis of a Law, Part 2," p. 39.

36. Schwartz, "Capsule Examination," pp. 297–300; Senzel, "Genesis of a Law, Part 2," p. 39.

37. 90 Stat. 2744; PL 94-579; 43 USC, secs. 1701–84; "Editorial: To Safeguard the Land," p. 38.

38. Reinhold, "Military and Preservationists Clash over Mojave's Future," p. 1; Baker, "Winning (and Losing) the West," pp. 56–60; *Wilderness* (Summer 1986) (BLM wilderness is the theme of the entire issue). Thus far, large-scale wilderness designation has only taken place in Arizona and California.

39. Coggins, "Law of Public Rangeland Management V," p. 507; Baker, "Frustration of FLPMA," p. 13; Culhane, *Public Lands Politics*; Miller, "FLPMA," p. 268.

40. Gregg as quoted in Baker, "Frustration of FLPMA," p. 13; ibid., p. 14; Coggins, "Law of Public Rangeland Management V," p. 507; Dana and Fairfax, *Forest and Range Policy*, p. 344; Durant, "Toward Assessing the Administrative Presidency"; Durant, *Administrative Presidency Revisited*.

41. Baker, "Frustration of FLPMA," p. 13; Dana and Fairfax, *Forest and Range Policy*, p. 341; Office of Management and Budget, *Budget of the United States, Fiscal Year 1988 (and Appendix)*; U.S. Department of the Interior, Bureau of Land Management, *Public Land Statistics 1987*; BLM, *Public Land Statistics 1984*; BLM, *Public Land Statistics 1978*. The general increases in BLM employees and budget reflects, to some degree, an increased responsibility and emphasis on minerals management under the Reagan administration. The decrease in lands managed by the BLM between 1978 and 1983 reflects adjustments due to the reclassification of lands in Alaska following the passage of the Alaska Lands Act in 1980. The BLM budget and employee figures for 1988 include the Office of Surface Mining and Regulation.

42. Coggins, "Law of Public Rangeland Management V," p. 507; Baker, "Frustration of FLPMA," p. 14.

43. On FLPMA and the Sagebrush Rebellion, see Hornblower, "Sagebrush Revolution," p. B1; Stegner, "Land: America's History Teacher," pp. 5–13; on FLPMA and the legislative veto, see Baker, "Chadha and the Public Lands"; Gaetke, "Separation of Powers"; Lee, "FLPMA's Legislative Veto Provisions and *INS v. Chadha*"; Sullivan, "Power of Congress under the Property Clause."

44. Secretary of the Interior and Secretary of Agriculture, *Study of Fees for Grazing Livestock on Federal Lands*, pp. 2-4, 2-5, 2-29; Williamson, "Where the Grass Is Greenest," pp. 30–31. For a general discussion of grazing fees on Grazing Service/BLM lands prior to 1970, see Calef, *Private Grazing and Public Lands*, pp. 72–76; Clawson, *Bureau of Land Management*, pp. 170–77; Foss, *Politics and Grass*, pp. 171–93; Nielsen, "Grazing Fees for Public Lands"; Nielsen and Wennergren, "Public Policy and Grazing Fees on Federal Lands"; Secretary of the Interior and Secretary of Agriculture, *Study of Fees for Grazing Livestock on Federal Lands*, pp. 2-2–2-9.

45. Foss, *Politics and Grass*, pp. 171–93.

46. Rowley, *U.S. Forest Service Grazing and Rangelands*, pp. 241–43; Secretary of the Interior and Secretary of Agriculture, *Study of Fees for Grazing Livestock on Federal Lands*, pp. 2-11–2-14.

47. Secretary of the Interior and Secretary of Agriculture, *Study of Fees for Grazing Livestock on Federal Lands*, pp. 2-17–2-23.

48. Ibid., p. 2-27.

49. "Editorial: The Grazing Land Must Be Restored," pp. 6–7; Secretary of the Interior and Secretary of Agriculture, *Study of Fees for Grazing Livestock on Federal Lands*, pp. 2-27–2-28.

50. *Pankey Land and Cattle Company v. Hardin*, 427 F. 2d 43 (1970); *Broadbent v. Hickel* (1969).

51. The difference between the value of the private land alone and the private land with associated permits was the permit value, the existence of which has been documented by numerous studies. A study of sales and bank loans in New Mexico revealed that the value of BLM permits ranged from $667 to $888 per AUM. Another study, however, argued that a permit value no longer existed due to increased fees and increased uncertainty; hence "there may not be a *stream of expected returns . . .* to be capitalized into the permit value." See Fowler and Gray, "Market Values of Federal Grazing Permits in New Mexico," p. 112; Martin, "Distribution of Benefits and Costs Associated with Public Rangelands," p. 252; Winter and Whittaker, "Relationship Between Private Ranchland Prices and Public-Land Grazing Permits," p. 420. For general discussions of permit value, see Calef, *Private Grazing and Public Lands*, pp. 272–73; Coggins and Lindeberg-Johnson, "Law of Public Rangeland Management II," pp. 74–75; Gardner, "Transfer Restrictions and Misallocation in Grazing Public Range"; Nielsen, "Grazing Fees for Public Lands," pp. 3–4; Reavley, "Our Point of View," pp. 40–41; Roberts, "Economic Foundations for Grazing Use Fees," p. 726.

52. See Libecap, *Locking Up the Range*.

53. The senators were Gordon Allott (R, Colo.), Henry Bellman (R, Okla.), Alan Bible (D, Nev.), Frank Church (D, Idaho), Peter Dominick (R, Colo.), Clifford Hansen (R, Wyo.), Mark Hatfield (R, Ore.), Len Jordan (R, Idaho), and Gale McGee (R, Wyo.). See Frome, "President's Environmental Crusade and the Public Lands," pp. 36–38.

54. Secretary of the Interior and Secretary of Agriculture, *Study of Fees for Grazing Livestock on Federal Lands*, pp. 2-28–2-31; Williamson, "Where the Grass Is Greenest," pp. 30–31.

55. As quoted in Williamson, "Where the Grass Is Greenest," pp. 30–31; Secretary of the Interior and Secretary of Agriculture, *Study of Fees for Grazing Livestock on Federal Lands*, pp. 2-28–2-31.

56. "Grazing Fees Boost Due in '78 Trimmed by Carter Proposal," p. 38; "Interior Proposes Boost in Fees for Livestock Grazing," p. A22.

57. Secretary of the Interior and Secretary of Agriculture, *Study of Fees for Grazing Livestock on Federal Lands*, pp. 4-1–4-31. The modified version of the current system included improved data collection on private lease rates, increasing rates 25

percent annually until FMV was realized, and limiting future charges to 12 percent per year once FMV was achieved. The National Cattlemen's Association formula had FMV equaling the Beef Price Index minus the Prices Paid Index (a cost of production index) plus 100 (this was termed the Combined Index). The House Interior Committee formula was almost identical to this. The technical committee proposed a formula of FMV equal to $1.23 (the 1966 base rate) times [private grazing land index plus combined index] divided by 100. The AFBF proposal subtracted the permit value from the $1.23, then duplicated the technical committee report.

58. "Public Grazing Land Laws," pp. 716–18; "Federal Grazing Fees Are Unchanged for Now," p. 48.

59. 92 Stat. 1803; PL 95-514; 43 USC, secs. 1901–8. The technical committee grazing fee formula is: FMV = $1.23 × (Forage Value Index + Combined Index) ÷ 100. The Combined Index = Beef Cattle Price Index − Price Paid Index. Coggins, "Law of Public Rangeland Management IV"; "Public Grazing Land Laws," pp. 716–18; Cox, "Deterioration of Southern Arizona's Grasslands."

60. Abbey, "Even the Bad Guys Wear White Hats," pp. 52–53. For example, Denzel and Nancy Ferguson argue that "the ridiculously low cost of public forage invites overgrazing and makes profitable the grazing of degraded public lands that could not support grazing in a free-market economy. Thus, the current system subsidizes and perpetuates operators who mine a subsistence living from public rangelands, but who would not survive in a truly competitive market. In short, large members of uneconomical ranches are being kept afloat by federal largesse" (*Sacred Cows at the Public Trough*, p. 202).

61. "Grazing Fee for Cattle on U.S. Land Is Boosted," p. 36; Williamson, "Where the Grass Is Greenest," pp. 30–31.

62. "Commission Offers Ways to Cut Costs," p. B7; Schmidt, "U.S. Easing Control Over Western Rangeland," p. 1; Zaleha, "Rise and Fall of BLM's 'Cooperative Management Agreements.'"

63. Gregg as quoted in Schmidt, "U.S. Easing Control Over Western Rangeland," p. 10; Anderson, "Leasing to Graze on U.S. Range Beats Rustling," p. F21; "Federal Agency Seeks to Bar Leasing of Land for Grazing," p. 33; Zaleha, "Rise and Fall of BLM's 'Cooperative Management Agreements.'"

64. Grant, "U.S. Eyes Higher Grazing Fees," p. A25.

65. National Wildlife Federation and Wilderness Society as quoted in Peterson, "Home on the Range," p. 4; Fradkin, "The Eating of the West," p. 94; "Grazing at the Bottom Line," pp. 6–7; Williamson, "Where the Grass Is Greenest," p. 30.

66. Fradkin, "The Eating of the West," p. 120; "Editorial: Government Grass," p. A14.

67. Borchers, "Reforming Federal Grazing Law"; Davis, "Cattlemen vs. Environmentalists," p. 1676.

68. As quoted in "Legislative Log: Seiberling Cosponsors Competitive Bid Grazing Fee Issue," p. 132; Borchers, "Reforming Federal Grazing Law"; Peterson, "Home on the Range."

69. As quoted in Stanfield, "Cowboys and Conservationists in Range War," pp.

1623–25; Council on Environmental Quality, *Environmental Quality 1984*, p. 259; Davis, "Cattlemen vs. Environmentalists," p. 1678; Nielsen, "Grazing Fees for Public Lands," p. 4; Williamson, "Where the Grass Is Greenest," p. 31.

70. "Legislative Log: Seiberling Cosponsors Competitive Bid Grazing Fee Issue," p. 132; "Legislative Log: Rangeland Bill Circulating," pp. 180–81.

71. Davis, "Cattlemen vs. Environmentalists," pp. 1676–78; "Legislative Log: Rangeland Bill Circulating," pp. 180–81; Williamson, "Range Is a Terrible Thing to Waste," pp. 30–33.

72. Garn as quoted in Davis, "Cattlemen vs. Environmentalists," p. 1677; "Legislative Log: Grazing Fee Formula Sticking Point in New Omnibus Range Bill," p. 218; "Legislative Log: Conservationists Say No to New Rangeland Bill," p. 271.

73. Reid, "Western Grazing Fees Slip Out of Budget Noose," p. A3; Shabecoff, "Rise in Federal Grazing Fees," p. 5.

74. Miller as quoted in Peterson, "OMB Urges Freezing Fees for Grazing Federal Land," p. A4; as quoted in "Wilderness Watch: Grazing Subsidy Survives," p. 4.

75. "Editorial: User Fees," p. A14.

76. As quoted in Shabecoff, "President Extends Grazing Fees," p. 10; "Low Grazing Fee Extended," p. A10.

77. As quoted in "Wilderness Watch: A 'Subsidy of Destruction' Brings Forth a Suit," p. 2; "Capital Corral: Fees!," p. 185; "Capital Corral: House Appropriations Subcommittee," p. 242. The nine groups bringing suit were: American Fisheries Society; California Trout, Inc.; Izaak Walton League; National Audubon Society; National Wildlife Federation; NRDC; Oregon Trout, Inc.; Sierra Club; and Wilderness Society. Nothing came of this suit, however.

Chapter Six

1. *United States v. Gratiot* (39 U.S. [14 Peters] 526 [1840]); *United States v. Gear* (44 U.S. [3 Howard] 120 [1845]); *U.S. v. Grimaud* (220 U.S. 506 [1911]); *Light v. U.S.* (220 U.S. 523 [1911]).

2. Gottlieb, *Wise Use Agenda*, pp. 5–18.

3. Baum, "Wise Guise," pp. 70–73, 90–93; Deal, *Greenpeace Guide to Anti-Environmental Organizations*.

4. The Report of the National Commission on the Environment, *Choosing a Sustainable Future*, p. xi.

5. Botkin, *Discordant Harmonies*, pp. 155–56 and more generally; Noss and Cooperrider, *Saving Nature's Legacy*, p. 391.

6. Durbin, "Ambitious Ecosystem Management Advances East," pp. 8–12; Kessler et al., "New Perspectives for Sustainable Natural Resources Management."

7. Drabelle, "'Obey the Law and Tell the Truth,'" pp. 29–33.

8. Brown and Harris, "U.S. Forest Service"; DeBonis, "Association of Forest Service Employees for Environmental Ethics," p. 14; *Inner Voice* (the newsletter of AFSEEE).

9. Brown and Harris, "United States Forest Service"; Brown and Harris, "Implications of Work Force Diversification."

10. Devall and Sessions, *Deep Ecology*; Foreman, *Confessions of an Eco-Warrior*; Manes, *Green Rage*; Nash, *Rights of Nature*, pp. 189–97; Scarce, *Eco-Warriors*.

11. Foreman, "Northern Rockies Ecosystem Protection Act," pp. 57–62; Rauber, "Priorities," pp. 40–42.

12. "Art, Owls, Oil Drilling Argued in Interior Bill," pp. 870–76; "Moratorium on Mining Claims Defeated," pp. 217–18.

13. "Mining Law Overhaul Is Stymied Again," pp. 282–85.

14. Benenson, "House Easily Passes Overhaul of 1872 Mining Law," pp. 3191–92; "Overhaul of Mining Law Advances," p. 262; Cushman, "Congress Drops Effort to Curb Public-Land Mining," p. A1.

15. Yaffe, *Wisdom of the Spotted Owl*, pp. 3–35.

16. Ibid., pp. 36–57.

17. Ibid., pp. 58–82.

18. Ibid., pp. 83–114.

19. Ibid., pp. 115–36.

20. "Option 9 Survives," p. 6; Watkins, "Perils of Option 9," pp. 6–9; Yaffe, *Wisdom of the Spotted Owl*, pp. 132–51.

21. We can't generalize about all institutions and institutional change based on these cases, however. Different policy regimes and different paths of development make statewide, ahistorical generalizations problematic at best.

22. Moe, "Politics of Bureaucratic Structure."

BIBLIOGRAPHY

Abbey, Edward. "Even the Bad Guys Wear White Hats." *Harper's* (January 1986): 51–55.

"Agency Rejects Industry Role." *New York Times,* January 28, 1987, p. A20.

"Agency Seeks to Sell 10% of National Forests." *Wall Street Journal,* August 11, 1982, p. 36.

"Agency Weighing Sale of 3.2% of Forest System." *New York Times,* March 16, 1983, p. 16.

Alderson, George. "Capitol Watch." *Living Wilderness* (October/December 1975): 41.

Allin, Craig W. *The Politics of Wilderness Preservation.* Westport, Conn.: Greenwood Press, 1982.

Alston, Richard M. *The Individual and the Public Interest: Political Ideology and National Forest Policy.* Boulder, Colo.: Westview Press, 1983.

Anderson, Ewan W. *Strategic Minerals: The Geopolitical Problems for the United States.* New York: Praeger, 1988.

———. *The Structure and Dynamics of U.S. Government Policymaking: The Case of Strategic Minerals.* New York: Praeger, 1988.

Anderson, Jack. "Leasing to Graze on U.S. Range Beats Rustling." *Washington Post,* March 31, 1984, p. F21.

"Art, Owls, Oil Drilling Argued in Interior Bill." *CQ Almanac* 46 (1990): 870–76.

Baden, John, and Dean Lueck. "A Property Rights Approach to Wilderness Management." In *Public Lands and the U.S. Economy: Balancing Conservation and Development,* edited by George M. Johnston and Peter M. Emerson, pp. 29–67. Boulder, Colo.: Westview Press, 1984.

Baker, James. "The Frustration of FLPMA." *Wilderness* (Winter 1983): 12–15, 22–24.

———. "Winning (and Losing) the West." *Sierra* (May/June 1985): 56–60.

Baker, Richard A. "The Conservation Congress of Anderson and Aspinall, 1963–64." *Journal of Forest History* 29 (1985): 104–19.

Baker, Timothy R. "Chadha and the Public Lands: Is FLPMA Affected?" *Public Land Law Review* 5 (1984): 55–67.

Bates, J. Leonard. "Fulfilling American Democracy: The Conservation Movement, 1907 to 1921." *Mississippi Valley Historical Review* 44 (1957–58): 29–57.

"Battle in the Finger Lakes: U.S. May Put a 13,000 Acre Haven Up for Grabs." *New York Times,* February 20, 1983, p. 50.

Baum, Dan. "Wise Guise." *Sierra* (May/June 1991): 70–73, 92–93.

Behan, R. W. "Forestry and the End of Innocence." *American Forests* (May 1975): 16–19, 38–49.

———. "The Myth of the Omnipotent Forester." *Journal of Forestry* 64 (1966): 398–407.

Benenson, Bob. "House Easily Passes Overhaul of 1872 Mining Law." *CQ Weekly Report* 51 (November 20, 1993): 3191–92.

Bennett, Douglas C., and Kenneth E. Sharpe. *Transnational Corporations versus the State: The Political Economy of the Mexican Auto Industry*. Princeton, N.J.: Princeton University Press, 1985.

Berry, Phillip. "An Analysis: The Public Land Law Review Commission Report." *Sierra Club Bulletin* (October 1970): 18–20.

Borchers, Timothy K. "Reforming Federal Grazing Law: Will Congress Pass Needed Legislation before the Cows Come Home?" *Journal of Legislation* 13 (1986): 216–41.

Botkin, Daniel B. *Discordant Harmonies: A New Ecology for the Twenty-first Century*. New York: Oxford University Press, 1990.

Bromley, Daniel W. "Public and Private Interests in the Federal Lands: Toward Conciliation." In *Public Lands and the U.S. Economy: Balancing Conservation and Development*, edited by George M. Johnston and Peter M. Emerson, pp. 3–18. Boulder, Colo.: Westview Press, 1984.

Brown, Greg, and Charles C. Harris. "The Implications of Work Force Diversification in the U.S. Forest Service." *Administration and Society* 25 (1993): 85–113.

———. "The United States Forest Service: Changing the Guard." *Natural Resources Journal* 32 (1992): 449–66.

———. "The U.S. Forest Service: Toward the New Resource Management Paradigm?" *Society and Natural Resources* 5 (1992): 231–45.

Calef, Wesley. *Private Grazing and Public Lands: Studies of the Local Management of the Taylor Grazing Act*. Chicago: University of Chicago Press, 1960.

Cameron, Eugene N. *At the Crossroads: The Mineral Problems of the United States*. New York: Wiley, 1986.

Camia, Catalina. "Administration Aims to Increase Grazing Fees, Tighten Rules." *CQ Weekly Report* 51 (August 14, 1993): 2223.

———. "The Filibuster Ends; Bill Clears; Babbitt Can Still Raise Fees." *CQ Weekly Report* 51 (November 13, 1993): 3112–13.

"Capital Corral: Fees!" *Rangelands* 8 (1986): 185.

"Capital Corral: House Appropriations Subcommittee." *Rangelands* 8 (1986): 242.

Carter, Luther J. "Minerals and Mining: Major Review of Federal Policy Is in Prospect." *Science* 198 (November 25, 1977): 809–11.

Chapman, H. H. "Reorganization and the Forest Service." *Journal of Forestry* 35 (1937): 427–38.

Clarke, Jeanne N., and Daniel McCool. *Staking Out the Terrain: Power Differentials among Natural Resource Management Agencies*. Albany: State University Press of New York, 1985.

Clary, David A. *Timber and the Forest Service*. Lawrence: University Press of Kansas, 1986.

Clawson, Marion. *The Bureau of Land Management*. New York: Praeger, 1971.

———. *The Federal Lands Revisited*. Baltimore, Md.: Johns Hopkins University Press for Resources for the Future, 1983.

Clayton, Richard D. "The Sagebrush Rebellion: Who Should Control the Public Lands?" *Utah Law Review* 1980:505–33.

Clepper, Henry. *Professional Forestry in the United States*. Baltimore, Md.: Johns Hopkins University Press for Resources for the Future, 1971.

Coggins, George C. "The Law of Public Rangeland Management III: A Survey of Creeping Regulation at the Periphery, 1934–1982." *Environmental Law* 13 (1983): 295–365.

———. "The Law of Public Rangeland Management IV: FLPMA, PRIA, and the Multiple Use Mandate." *Environmental Law* 14 (1983): 1–131.

———. "The Law of Public Rangeland Management V: Prescriptions for Reform." *Environmental Law* 14 (1984): 497–546.

Coggins, George C., and Margaret Lindeberg-Johnson. "The Law of Public Rangeland Management II: The Commons and the Taylor Act." *Environmental Law* 13 (1982): 1–101.

Coggins, George C., Parthenia B. Evans, and Margaret Lindeberg-Johnson. "The Law of Public Rangeland Management I: The Extent and Distribution of Federal Power." *Environmental Law* 12 (1982): 535–621.

Cohen, Michael P. *The History of the Sierra Club, 1892–1970.* San Francisco: Sierra Club Books, 1988.

"Commission Offers Ways to Cut Costs." *New York Times,* April 19, 1983, p. B7.

"Controversy over Wilderness Area Minerals Policy." *Congressional Digest* 61 (December 1982): 288–314.

Cook, James. "The Crisis That Didn't Happen." *Forbes,* November 22, 1982, pp. 91–94.

Council on Environmental Quality. *Environmental Quality: 22nd Annual Report of the Council on Environmental Quality.* Washington, D.C.: Government Printing Office, 1992.

Cox, David T. "Deterioration of Southern Arizona's Grasslands: Effects of New Federal Legislation Concerning Public Grazing Lands." *Arizona Law Review* 20 (1979): 697–741.

Culhane, Paul J. *Public Lands Politics: Interest Group Influence on the Forest Service and the Bureau of Land Management.* Baltimore, Md.: Johns Hopkins University Press for Resources for the Future, 1981.

Curlin, James W. "The Political Dimensions of Strategic Minerals." In *American Strategic Minerals,* edited by Gerard J. Mangone, pp. 135–47. New York: Crane Russak, 1984.

Cushman, John H. "Congress Drops Effort to Curb Public-Land Mining." *New York Times,* September 30, 1994, p. A1.

———. "A Consensus Approach on Land Use." *New York Times,* February 19, 1994, p. 8.

———. "Grazing Fees Cut from Senate Bill." *New York Times,* November 10, 1993, p. A20.

Dana, Samuel T. *Forest and Range Policy.* New York: McGraw-Hill, 1956.

Dana, Samuel T., and Sally K. Fairfax. *Forest and Range Policy.* 2d ed. New York: McGraw-Hill, 1980.

Davis, Joseph A. "Cattlemen vs. Environmentalists: Congress Tries for Compromise on Disputed Grazing Fee Issue." *CQ Weekly Report* 43 (August 24, 1985): 1676–78.

Davis, Tony. "Babbitt Cedes Grazing Reform to Congress." *High Country News,* January 23, 1995, p. 6.

Deacon, Robert T., and M. Bruce Johnson, eds. *Forestlands: Public and Private*. San Francisco: Pacific Institute for Public Policy Research, 1985.

Deal, Carl. *The Greenpeace Guide to Anti-Environmental Organizations*. Berkeley, Calif.: Odonian Press, 1993.

DeBonis, Jeff. "Association of Forest Service Employees for Environmental Ethics." *Wild Earth* (Summer 1991): 14.

Dennen, R. Taylor. "Cattlemen's Associations and Property Rights in Land in the American West." *Explorations in Economic History* 13 (1976): 423–36.

Derthick, Martha, and Paul J. Quirk. *The Politics of Deregulation*. Washington, D.C.: Brookings Institution Press, 1985.

Devall, Bill, and George Sessions. *Deep Ecology: Living as if Nature Mattered*. Salt Lake City: Peregrine Smith Books, 1985.

"Dig It." *Forbes*, November 1, 1975, p. 84.

Dowdle, Barney. "The Case for Privatizing Government Owned Timberlands." In *Private Rights and Public Lands*, edited by Phillip N. Truluck, pp. 71–83. Washington, D.C.: Heritage Foundation, 1983.

Drabelle, Dennis. "'Obey the Law and Tell the Truth': An Interview with Forest Service Chief Jack Ward Thomas." *Wilderness* (Fall 1994): 29–33.

Dupree, A. Hunter. *Science in the Federal Government*. 1957. Reprint, Baltimore, Md.: Johns Hopkins University Press, 1986.

Durant, Robert F. *The Administrative Presidency Revisited: Public Lands, the BLM, and the Regan Revolution*. Albany: State University of New York Press, 1992.

———. "Toward Assessing the Administrative Presidency: Public Lands, the BLM, and the Reagan Administration." *Public Administration Review* 47 (1987): 180–89.

Durbin, Kathie. "Ambitious Ecosystem Management Advances East." *High Country News*, September 19, 1994, pp. 8–12.

Eckes, Alfred E. *The United States and the Global Struggle for Minerals*. Austin: University of Texas Press, 1979.

"Editorial: Forever Wild." *Living Wilderness* (Summer/Fall 1957): ii.

"Editorial: Government Grass." *Washington Post*, July 23, 1985, p. A14.

"Editorial: The Grazing Land Must Be Restored." *American Forests* (January 1969): 6–7.

"Editorial: The Minerals Problem Is Not a Crisis." *New York Times*, August 7, 1981, p. A22.

"Editorial: Scuttling Land Reform." *New York Times*, December 17, 1973, p. 36.

"Editorial: To Safeguard the Land." *New York Times*, October 21, 1976, p. 38.

"Editorial: User Fees." *Washington Post*, February 3, 1986, p. A14.

Egan, Timothy. "Wingtip 'Cowboys' in Last Stand to Hold on to Low Grazing Fees." *New York Times*, October 29, 1993, p. A1.

Ellison, Joseph. "The Mineral Land Question in California, 1848–1866." In *The Public Lands*, edited by Vernon Carstensen, pp. 71–92. 1926. Reprint, Madison: University of Wisconsin Press, 1963.

Evans, Brock. "Washington Report: The Forgotten Public Domain." *Sierra Club Bulletin* (March 1976): 22–23.

Evans, Peter B., Dietrich Rueschemeyer, and Theda Skocpol, eds. *Bringing the State Back In*. New York: Cambridge University Press, 1985.

"Federal Agency Seeks to Bar Leasing of Land for Grazing." *New York Times*, March 17, 1985, sec. 1, p. 33.

"Federal Grazing Fees Are Unchanged for Now." *Wall Street Journal*, February 7, 1978, p. 48.

"Federal Land Urged as Strategic Mineral Source." *Aviation Week and Space Technology*, March 16, 1981, pp. 50–51.

Ferguson, Denzel, and Nancy Ferguson. *Sacred Cows at the Public Trough*. Bend, Ore.: Maverick Publications, 1983.

Flawn, Peter T. "Impact of Environmental Concerns on the Minerals Industry, 1975–2000." In *The Mineral Position of the United States, 1975–2000*, edited by Eugene N. Cameron, pp. 95–108. Madison: University of Wisconsin Press, 1973.

Foreman, Dave. *Confessions of an Eco-Warrior*. New York: Harmony Books, 1991.

———. "The Northern Rockies Ecosystem Protection Act and the Evolving Wilderness Area Model." *Wild Earth* (Winter 1993–94): 57–62.

Foss, Phillip O. *Politics and Grass*. Seattle: University of Washington Press, 1960.

Fowler, John M., and James R. Gray. "Market Values of Federal Grazing Permits in New Mexico." *Rangelands* 2 (1980): 112.

Fox, Stephen. *The American Conservation Movement: John Muir and His Legacy*. Madison: University of Wisconsin Press, 1985.

———. "We Want No Straddlers." *Wilderness* (Winter 1984): 4–19.

Fradkin, Philip L. "The Eating of the West." *Audubon* (January 1979): 94–121.

Friedman, Milton. *Capitalism and Freedom*. Chicago: University of Chicago Press, 1962.

Frome, Michael. *Battle for the Wilderness*. New York: Praeger, 1974.

———. *The Forest Service*. New York: Praeger, 1971.

———. "The President's Environmental Crusade and the Public Lands." *Field and Stream* (February 1970): 36–38, 141.

Gaetke, Eugene R. "Separation of Powers, Legislative Vetoes, and the Public Lands." *University of Colorado Law Review* 56 (1985): 559–80.

Gardner, B. Delworth. "Transfer Restrictions and Misallocation in Grazing Public Range." *Journal of Farm Economics* 44 (1962): 50–63.

Gates, Paul W. *History of Public Land Law Development*. Washington, D.C.: Government Printing Office, 1968.

———. "Pressure Groups and Recent American Land Policies." *Agricultural History* 55 (1981): 103–27.

Glover, James. *Wilderness Original: The Life of Bob Marshall*. Seattle: Mountaineers, 1986.

Goldwater, Barry. "U.S. Dependency on Foreign Sources for Critical Material." *Vital Speeches of the Day* 47 (1981): 517–20.

Gottlieb, Alan M., ed. *The Wise Use Agenda: The Citizen's Guide to Environmental Resource Issues*. Bellevue, Wash.: Free Enterprise Press, 1989.

Graf, William L. *Wilderness Preservation and the Sagebrush Rebellions*. Savage, Md.: Rowman and Littlefield, 1990.

Grant, Bob. "U.S. Eyes Higher Grazing Fees." *Washington Post*, April 19, 1985, p. A25.

"Grazing at the Bottom Line." *Wilderness* (Fall 1985): 6–7.

"Grazing Fee for Cattle on U.S. Land Is Boosted." *Wall Street Journal*, January 10, 1979, p. 36.

"Grazing Fees Boost Due in '78 Trimmed by Carter Proposal." *Wall Street Journal*, October 24, 1977, p. 38.

Greever, William S. *The Bonanza West: The Story of Western Mining Rushes, 1848–1900.* Norman: University of Oklahoma Press, 1963.

Hackett, David. "Wallop Proposes Grazing Fee Increase." *Casper Star-Tribune*, July 31, 1993, p. A1.

Hagenstein, Perry R. "The Federal Lands Today—Uses and Limits." In *Rethinking the Federal Lands*, edited by Sterling Brubaker, pp. 74–107. Baltimore, Md.: Johns Hopkins University Press for Resources for the Future, 1984.

Hall, Peter A. "Conclusion: The Politics of Keynesian Ideas." In *The Political Power of Economic Ideas: Keynesianism across Nations*, edited by Peter A. Hall, pp. 361–91. Princeton, N.J.: Princeton University Press, 1989.

Hammer, Alexander R. "People and Business." *New York Times*, October 17, 1975, p. 55.

Hanke, Steve H. "Grazing for Dollars." *Reason* 14 (July 1982): 43–44.

———. "On Privatizing the Public Domain." In *Private Rights and Public Lands*, edited by Phillip N. Truluck, pp. 85–88. Washington, D.C.: Heritage Foundation, 1983.

———. "The Privatization Debate: An Insider's View." *Cato Journal* 2 (1982): 653–62.

———. "Privatize Those Lands." *Reason* 13 (March 1982): 39.

———. "Wise Use of Federal Land." *New York Times*, May 6, 1983, p. 31.

Harmon, Frank J. "What Should Foresters Wear?: The Forest Service's Seventy-Five-Year Search for a Uniform." *Journal of Forest History* 24 (1980): 188–99.

Hays, Samuel P. *Conservation and the Gospel of Efficiency: The Progressive Conservation Movement, 1890–1920.* 1959. Reprint, New York: Atheneum, 1975.

Hendricks, Rickey L. "Farrington R. Carpenter: New Deal Cowboy." *Midwest Review*, 2d ser., 5 (1983): 20–36.

Hershey, Robert D. "New Senate Energy Chief Sees No 'Radical Change' on Issues." *New York Times*, November 26, 1980, p. D6.

Hershiser, Beulah. "The Influence of Nevada on the National Mining Legislation of 1866." *Nevada Historical Society, Third Biennial Report* (1911–12): 127–67.

Hession, Jack M. "The Legislative History of the Wilderness Act." Master's thesis, San Diego State College, 1967.

Hill, Gladwin. "Revised Policy for U.S. Lands Asked in Study." *New York Times*, June 24, 1970, p. 1.

Hornblower, Margot. "The Sagebrush Revolution." *Washington Post*, November 11, 1979, p. B1.

Hurst, James Willard. *Law and the Conditions of Freedom in the Nineteenth Century United States.* Madison: University of Wisconsin Press, 1964.

"Interior Dept. Seeks Mining Permits in Parks." *New York Times*, July 17, 1981, p. A10.

"Interior Proposes Boost in Fees for Livestock Grazing." *Washington Post*, October 22, 1977, p. A22.

Ise, John. *The United States Forest Policy*. New Haven, Conn.: Yale University Press, 1920.

Kaufman, Herbert. *The Forest Ranger: A Study in Administrative Behavior*. Baltimore, Md.: Johns Hopkins University Press for Resources for the Future, 1960.

Kessler, Winifred B., Hal Salwasser, Charles W. Cartwright, and James A. Caplan. "New Perspectives for Sustainable Natural Resources Management." *Ecological Applications* 2 (1992): 221–25.

Kingery, Hugh E. "The Public Grazing Lands." *Denver Law Journal* 43 (1966): 329–48.

Klemme, Marvin. *Home Rule on the Range: Early Days of the Grazing Service*. New York: Vantage Press, 1984.

Klyza, Christopher McGrory. "A Window of Autonomy: State Autonomy and the Forest Service in the Early 1900s." *Polity* 25 (1992): 173–96.

Knous, William L. "The Use of Public Lands—a National Problem." *State Government* 20 (August 1947): 209–12.

"Knowledge, Power and International Policy Coordination." *International Organization* 46 (Winter 1992): 1–390.

Kolko, Gabriel. *The Triumph of Conservatism: A Reinterpretation of American History, 1900–1916*. New York: Free Press, 1963.

Krauss, Clifford. "Clinton Planning to Increase Fees on Grazing Lands." *New York Times*, August 10, 1993, p. A1.

Kreiger, Martin H. "What's Wrong with Plastic Trees?" *Science* 179 (February 2, 1973): 446–55.

"Land Sale Goal Is Scaled Back." *Washington Post*, February 2, 1983, p. A21.

Lavender, David. "Decision for Permanence." *Wilderness* (Winter 1983): 4–11.

Lee, William P. "FLPMA's Legislative Veto Provisions and *INS v. Chadha*: Who Controls the Federal Lands?" *Boston College Environmental Affairs Law Review* 12 (1985): 791–821.

"Legislative Log: Conservationists Say No to New Rangeland Bill." *Rangelands* 7 (1985): 271.

"Legislative Log: Grazing Fee Formula Sticking Point in New Omnibus Range Bill." *Rangelands* 7 (1985): 218.

"Legislative Log: Rangeland Bill Circulating." *Rangelands* 7 (1985): 180–81.

"Legislative Log: Seiberling Cosponsors Competitive Bid Grazing Fee Issue." *Rangelands* 7 (1985): 132.

Leopold, Aldo. *A Sand County Almanac*. 1949. Reprint, New York: Oxford University Press, 1966.

———. "The Wilderness and Its Place in Forest Recreational Policy." In *The River of the Mother of God and Other Essays*, pp. 78–81. 1921. Reprint, Madison: University of Wisconsin Press, 1991.

Leshy, John D. *The Mining Law: A Study in Perpetual Motion*. Baltimore, Md.: Johns Hopkins University Press for Resources for the Future, 1987.

———. "Sharing Federal Multiple Use Lands—Historic Lessons and Speculations for the Future." In *Rethinking the Federal Lands*, edited by Sterling Brubaker, pp. 235–74. Baltimore, Md.: Johns Hopkins University Press for Resources for the Future, 1984.

——. "Unraveling the Sagebrush Rebellion: Law, Politics and Federal Lands." *University of California, Davis Law Review* 14 (1980): 317–55.

Lewis, Bernard J. "The Reagan Administration's Federal Land Sales Program: Economic, Legal and Jurisdictional Issues." Staff Paper Series No. 37, University of Minnesota, Department of Forest Resources, 1983.

Libecap, Gary D. "Economic Variables and the Development of the Law: The Case of Western Mineral Rights." *Journal of Economic History* 38 (1978): 338–62.

——. *The Evolution of Private Mineral Rights: Nevada's Comstock Lode.* New York: Arno Press, 1978.

——. "Government Support of Private Claims to Public Minerals: Western Mineral Rights." *Business History Review* 53 (1979): 364–85.

——. *Locking Up the Range.* San Francisco: Pacific Institute for Public Policy Research, 1981.

"Low Grazing Fee Extended." *Washington Post*, February 15, 1986, p. A10.

Lowi, Theodore J. "American Business, Public Policy, Case-Studies, and Political Theory." *World Politics* 16 (1964): 676–715.

——. *The End of Liberalism.* 2d ed. New York: Norton, 1979.

Manes, Christopher. *Green Rage: Radical Environmentalism and the Unmaking of Civilization.* Boston: Little, Brown, 1990.

March, James G., and Johan P. Olsen. "The New Institutionalism: Organizational Factors in Political Life." *American Political Science Review* 78 (1984): 734–49.

Marsh, S. P., S. J. Kropschot, and R. G. Dickinson. *Wilderness Potential: Assessment of Mineral-Resource Potential in U.S. Forest Service Lands Studied, 1964–1984, vols. 1 and 2.* Geological Survey Professional Paper 1300. Washington, D.C.: Government Printing Office, 1984.

Marshall, Robert. "The Problem of Wilderness." In *The American Environment: Readings in the History of Conservation*, 2d ed., edited by Roderick Nash, pp. 121–26. 1930. Reprint, Reading, Pa.: Addison-Wesley, 1976.

Martin, William E. "The Distribution of Benefits and Costs Associated with Public Rangelands." In *Public Lands and the U.S. Economy: Balancing Conservation and Development*, edited by George M. Johnston and Peter M. Emerson, pp. 229–56. Boulder, Colo.: Westview Press, 1984.

Marx, Leo. *The Machine in the Garden: Technology and the Pastoral Idea in America.* New York: Oxford University Press, 1964.

Mayer, Carl J., and George A. Riley. *Public Domain, Private Dominion: A History of Public Mineral Policy in America.* San Francisco: Sierra Club Books, 1985.

McCloskey, Michael. "An Analysis: The Public Land Law Review Commission Report." *Sierra Club Bulletin* (October 1970): 21–30.

——. "The Wilderness Act of 1964: Its Background and Meaning." *Oregon Law Review* 45 (1966): 288–321.

McConnell, Grant. "The Conservation Movement—Past and Present." *Western Political Quarterly* 7 (1954): 463–78.

——. *Private Power and American Democracy.* New York: Random House, 1966.

Meine, Curt. *Aldo Leopold: His Life and Work.* Madison: University of Wisconsin Press, 1988.

Mercure, Delbert V., and William M. Ross. "The Wilderness Act: A Product of Congressional Compromise." In *Congress and the Environment*, edited by Richard A. Cooley and Geoffrey Wandesforde-Smith, pp. 47–64. Seattle: University of Washington Press, 1970.

"Metals Found on Oil Reserve." *New York Times*, July 17, 1980, p. D6.

Meyer, Arthur B. "Forests and Forestry." In *Origins of American Conservation*, edited by Henry Clepper, pp. 38–56. New York: Ronald Press, 1966.

Mikesell, Raymond F. *Nonfuel Minerals: Foreign Dependence and National Security*. Ann Arbor: University of Michigan Press, 1987.

Miller, Richard O. "FLPMA: A Decade of Management under the BLM Organic Act." *Policy Studies Journal* 14 (1985): 265–73.

"A Minerals Cartel Would Be Worse than the Energy Crisis: Interview with the Secretary of the Interior Rogers Morton." *Forbes*, February 15, 1974, pp. 48–49.

"Mining Hopes in New Mexico." *New York Times*, April 12, 1980, p. 38.

"Mining Law Overhaul Is Stymied Again." *CQ Almanac* 48 (1992): 282–85.

Mitchell, John G. "In Wildness Was the Preservation of a Smile." *Wilderness* (Summer 1985): 10–21.

Moe, Terry M. "The Politics of Bureaucratic Structure." In *Can the Government Govern?*, edited by John E. Chubb and Paul E. Peterson, pp. 267–329. Washington, D.C.: Brookings Institution Press, 1989.

Mollison, Richard M., and Richard W. Eddy. "The Sagebrush Rebellion: A Simplistic Response to the Complex Problems of Federal Land Management." *Harvard Journal on Legislation* 19 (1982): 97–142.

"Moratorium on Mining Claims Defeated." *CQ Almanac* 47 (1991): 217–18.

"More Commercial Use of Public Land Urged by Congressional Panel in Review of Policy." *Wall Street Journal*, June 24, 1970, p. 5.

Nash, Roderick F. *The Rights of Nature: A History of Environmental Ethics*. Madison: University of Wisconsin Press, 1989.

———. *Wilderness and the American Mind*. 3d ed. New Haven, Conn.: Yale University Press, 1982.

Netschert, Bruce C. "Better Management of Nonfuel Minerals on Federal Land: A Look at the Issues." In *Public Lands and the U.S. Economy: Balancing Conservation and Development*, edited by George M. Johnston and Peter M. Emerson, pp. 189–204. Boulder, Colo.: Westview Press, 1984.

"News: BLM Organic Act Passes House." *Sierra Club Bulletin* (September 1976): 31.

"News: PLLRC." *Sierra Club Bulletin* (September 1971): 3–4.

Nielsen, Darwin B. "Grazing Fees for Public Lands: What's Fair?" *Utah Science* 43 (1982): 1–5.

Nielsen, Darwin B., and E. Boyd Wennergren. "Public Policy and Grazing Fees on Federal Lands: Some Unresolved Issues." *Land and Water Law Review* 5 (1970): 293–320.

"No Large Land Sales Planned, Watt Says." *Washington Post*, June 30, 1983, p. A18.

Nordlinger, Eric A. *On the Autonomy of the Democratic State*. Cambridge, Mass.: Harvard University Press, 1981.

Noss, Reed F., and Allen Y. Cooperrider. *Saving Nature's Legacy: Protecting and Restoring Biodiversity.* Washington, D.C.: Island Press, 1994.

"Now the Squeeze on Metals." *Business Week,* July 2, 1979, p. 50.

"NPCA at Work: BLM: Public Lands in Limbo." *National Parks* (December 1973): 28–29.

Oelschlaeger, Max. *The Idea of Wilderness.* New Haven, Conn.: Yale University Press, 1991.

Olson, Sherry H. *The Depletion Myth: A History of Railroad Use of Timber.* Cambridge, Mass.: Harvard University Press, 1971.

"Option 9 Survives." *High Country News,* January 23, 1995, p. 6.

"Overhaul of Mining Law Advances." *CQ Almanac* 49 (1993): 262.

Overton, J. Allen. "Letter: A Mining Industry Barred from Preparing for Its Nation's Needs." *New York Times,* September 7, 1981, p. A14.

———. "The Mining Industry: The Dangers and Stranglehold of Regulations." *Vital Speeches of the Day* 44 (1978): 253–56.

Palmer, Tim. *Endangered Rivers and the Conservation Movement.* Berkeley and Los Angeles: University of California Press, 1986.

Paul, Rodman W. *Mining Frontiers of the Far West, 1848–1880.* New York: Holt, Rinehart, and Winston, 1963.

Peffer, E. Louise. *The Closing of the Public Domain: Disposal and Reservation Policies, 1900–1950.* Stanford, Calif.: Stanford University Press, 1951.

Penny, J. Russell, and Marion Clawson. "Administration of Grazing Districts." In *The Public Lands,* edited by Vernon Carstensen, pp. 461–78. Madison: University of Wisconsin Press, 1963.

Peterson, Cass. "OMB Urges Freezing Fees for Grazing Federal Land." *Washington Post,* January 28, 1986, p. A4.

Peterson, Iver. "Home on the Range, But at What Price?" *New York Times,* October 27, 1985, sec. 4, p. 4.

Petulla, Joseph M. *American Environmentalism: Values, Tactics, Priorities.* College Station: Texas A&M University Press, 1980.

Pinchot, Gifford. *Breaking New Ground.* Seattle: University of Washington Press, 1947.

———. *The Fight for Conservation.* New York: Doubleday, Page, 1910.

Pinkett, Harold T. "The Forest Service, Trail Blazer in Recordkeeping Methods." *American Archivist* 22 (1959): 419–26.

Polenberg, Richard. "Conservation and Reorganization: The Forest Service Lobby, 1937–1938." *Agricultural History* 39 (1965): 230–39.

Ponder, Stephen. "Federal News Management in the Progressive Era: Gifford Pinchot and the Conservation Crusade." *Journalism History* 13, no. 2 (1986): 42–48.

———. "Gifford Pinchot: Press Agent for Forestry." *Journal of Forest History* 31 (1987): 26–35.

"Problems Found in Mineral Rights." *New York Times,* December 16, 1984, sec. 1, p. 33.

"Public Grazing Land Laws." *CQ Almanac* 34 (1978): 716–18.

"Public Land Management." *CQ Almanac* 32 (1976): 182–88.

Quirk, Paul J. "Deregulation and the Politics of Ideas in Congress." In *Beyond Self-Interest*, edited by Jane Mansbridge, pp. 183–99. Chicago: University of Chicago Press, 1990.

——. "In Defense of the Politics of Ideas." *Journal of Politics* 50 (1988): 31–41.

Raloff, J. "Watt Yields on Wilderness Leasing." *Science News* 123 (January 8, 1983): 21.

Rauber, Paul. "Priorities: The Last, Best Chance." *Sierra* (March/April 1994): 40–42.

Reavley, William L. "Our Point of View: Public Rights vs. Stockmen's Rights." *National Wildlife* (June/July 1969): 40–41.

Reich, Charles A. *Bureaucracy and the Forests*. Santa Barbara, Calif.: Center for the Study of Democratic Institutions, 1962.

Reid, T. R. "Western Grazing Fees Slip Out of Budget Noose." *Washington Post*, February 2, 1986, p. A3.

Reinhold, Robert. "Military and Preservationists Clash over Mojave's Future." *New York Times*, June 25, 1988, p. 1.

The Report of the National Commission on the Environment. *Choosing a Sustainable Future*. Washington, D.C.: Island Press, 1993.

Roberts, N. K. "Economic Foundations for Grazing Use Fees on Public Lands." *Journal of Farm Economics* 45 (1963): 721–31.

Roberts, Paul H. *Hoof Prints on Forest Ranges*. San Antonio, Tex.: Naylor, 1963.

Roosevelt, Theodore. *An Autobiography*. New York: Macmillan, 1913.

Roth, Dennis. "The National Forests and the Campaign for Wilderness Legislation." *Journal of Forest History* 28 (1984): 112–25.

Rowley, William D. *U.S. Forest Service Grazing and Rangelands: A History*. College Station: Texas A&M University Press, 1985.

Runge, Carlisle F. "An Economist's Critique of Privatization." In *Public Lands and the U.S. Economy: Balancing Conservation and Development*, edited by George M. Johnston and Peter M. Emerson, pp. 69–75. Boulder, Colo.: Westview Press, 1984.

——. "The Fallacy of 'Privatization.'" *Journal of Contemporary Studies* 7, no. 1 (1984): 3–17.

——. "Rejoinder: Looking Again at the New Resource Economics." *Journal of Contemporary Studies* 7, no. 2 (1984): 63–87.

Runte, Alfred. *National Parks: The American Experience*. 2d ed. Lincoln: University of Nebraska Press, 1987.

Sax, Joseph L. "The Claim for Retention of the Public Lands." In *Rethinking the Federal Lands*, edited by Sterling Brubaker, pp. 125–48. Baltimore, Md.: Johns Hopkins University Press for Resources for the Future, 1984.

——. "Why We Will Not (Should Not) Sell the Public Lands: Changing Conceptions of Private Property." *Utah Law Review* 1983:313–26.

Scarce, Rik. *Eco-Warriors: Understanding the Radical Environmental Movement*. Chicago: Noble, 1990.

Schiff, Ashley L. *Fire and Water: Scientific Heresy in the Forest Service*. Cambridge, Mass.: Harvard University Press, 1962.

———. "Innovation and Administrative Decision Making: The Conservation of Land Resources." *Administrative Science Quarterly* 11 (1966): 1–30.

Schmidt, William E. "Skeptics in West Hear Case for U.S. Land Sales." *New York Times*, March 21, 1983, p. 12.

———. "U.S. Easing Control over Western Rangeland." *New York Times*, February 14, 1983, p. 1.

Schneider, Keith. "House and Senate Agree to Raise Fees for Grazing on Federal Land." *New York Times*, October 8, 1993, p. A27.

———. "Senate Hands Clinton Setback on Grazing Fee." *New York Times*, September 16, 1993, p. A21.

Schwartz, Eleanor R. "A Capsule Examination of the Legislative History of the Federal Land Policy and Management Act of 1976." *Arizona Law Review* 21 (1979): 285–300.

Scott, Valerie W. "The Range Cattle Industry: Its Effect on Western Land Law." *Montana Law Review* 28 (1967): 155–83.

Senzel, Irving. "Genesis of a Law, Part 1." *American Forests* (January 1978): 30–32, 61–64.

———. "Genesis of a Law, Part 2." *American Forests* (February 1978): 32–39.

Shabecoff, Philip. "Bill Would Bar Wilds Drilling." *New York Times*, February 25, 1982, p. A22.

———. "Debate over Wilderness Area Leasing Intensifies." *New York Times*, February 15, 1982, p. D6.

———. "Drilling and Mining Planned in 5 Recreation Areas." *New York Times*, December 18, 1981, p. A33.

———. "Historic Battle over a Yosemite Lake Is Back." *New York Times*, August 6, 1987, p. A16.

———. "President Extends Grazing Fees; Conservationists Assail Decision." *New York Times*, February 17, 1986, sec. 1, p. 10.

———. "Reagan Tells of Plan to Open U.S. Lands for Minerals Search." *New York Times*, April 6, 1982, p. A1.

———. "Rise in Federal Grazing Fees Is Sought by Wildlife Group." *New York Times*, December 31, 1985, sec. 2, p. 5.

———. "U.S. Cuts Off Protection of Millions of Acres." *New York Times*, April 24, 1985, p. A1.

———. "Watt Removes Agency's Land From Sale Plan: He Is Said to See Program as a Political Liability." *New York Times*, July 28, 1983, p. 1.

———. "Watt to Seek Ban on Mineral Leases for Federal Lands." *New York Times*, February 22, 1982, p. A1.

Shaine, Benjamin A. "Alaska's Minerals." *Sierra* (July 1978): 27–28.

Shanks, Bernard. *This Land Is Yours: The Struggle to Save America's Public Lands*. San Francisco: Sierra Club Books, 1984.

Shinn, Charles H. "Land Laws of Mining Districts." In *Johns Hopkins University Studies in Historical and Political Science*, 2d ser., pp. 548–615. Baltimore, Md.: Johns Hopkins University Press, 1884.

———. *Mining Camps: A Case Study in American Frontier Government*. 1884. Reprint, New York: Harper and Row, 1965.

Short, C. Brant. *Ronald Reagan and the Public Lands: America's Conservation Debate, 1979–1984.* College Station: Texas A&M Press, 1989.

Simmons, Randy T., and John Baden. "The Theory of the NRE." *Journal of Contemporary Studies* 7, no. 2 (1984): 45–52.

"6 Million Acres of U.S. Forest Eyed for Sale." *Washington Post*, March 16, 1983, p. A8.

Skowronek, Stephen. *Building a New American State: The Expansion of National Administrative Capacities, 1877–1920.* New York: Cambridge University Press, 1982.

Slappey, Sterling G. "We're Headed for a Metals Crunch." *Nation's Business* (December 1974): 21–24.

Smith, Duane A. *Mining America: The Industry and the Environment, 1800–1980.* Lawrence: University Press of Kansas, 1987.

Smith, Henry N. *Virgin Land.* Cambridge, Mass.: Harvard University Press, 1950.

Society of American Foresters. *Forestry for America's Future, Proceedings of the SAF 1976 Convention.* Washington, D.C.: Society of American Foresters, 1977.

Stanfield, Rochelle L. "Cowboys and Conservationists in Range War over Grazing Fees on Public Lands." *National Journal* 17 (July 13, 1985): 1623–25.

Steen, Harold K. *The U.S. Forest Service: A History.* Seattle: University of Washington Press, 1976.

Stegner, Wallace. "Land: America's History Teacher." *Living Wilderness* (Summer 1981): 5–13.

Stoler, Peter. "Land Sale of the Century." *Time*, August 23, 1982, pp. 16–22.

Stout, Joe A. "Cattlemen, Conservationists, and the Taylor Grazing Act." *New Mexico Historical Review* 45 (1970): 311–32.

"The Strategic Minerals Fallacy." *Living Wilderness* (Summer 1982): 40–41.

Stroup, Richard L., and John A. Baden. "Endowment Areas: A Clearing in the Policy Wilderness?" *Cato Journal* 2 (1982): 691–708.

——. *Natural Resources: Bureaucratic Myths and Environmental Management.* San Francisco: Pacific Institute for Public Policy Research, 1983.

Struck, Myron. "Block Wants to Sell Forest Land to Chip away at Debt." *Washington Post*, August 12, 1982, p. A21.

——. "Sale of Surplus Lands a Bust in Its First Year." *Washington Post*, October 11, 1983, p. A13.

Sullivan, Roger M. "The Power of Congress under the Property Clause: A Potential Check on the Effect of the Chadha Decision on Public Land Legislation." *Public Land Law Review* 6 (1985): 65–102.

Sumner, David. "Wilderness and the Mining Law." *Living Wilderness* (Spring 1973): 8–18.

Sundquist, James L. *Politics and Policy: The Eisenhower, Kennedy, and Johnson Years.* Washington, D.C.: Brookings Institution Press, 1968.

"Surplus Western Lands May Still Be for Sale." *Washington Post*, July 29, 1983, p. A17.

Swenson, Robert W. "Legal Aspects of Mineral Resources Exploitation." In *History of Public Land Law Development*, by Paul W. Gates, pp. 699–764. Washington, D.C.: Government Printing Office, 1968.

"Symposium on the Public Land Law Review Commission." *Land and Water Law Review* 6 (1970).

"Symposium on the Public Land Law Review Commission Revisited." *Denver Law Journal* 54 (1977).

Taylor, Andrew. "President Will Not Use Budget to Rewrite Land-Use Laws." *CQ Weekly Report* 51 (April 3, 1993): 833–34.

Tilton, John E. *The Future of Nonfuel Minerals.* Washington, D.C.: Brookings Institution Press, 1977.

Tilton, John E., and Hans H. Landsberg. "Nonfuel Minerals: The Fear of Shortages and the Search for Policies." In *U.S. Interests and Global Natural Resources: Energy, Minerals, Food,* edited by Emery N. Castle and Kent A. Price, pp. 48–80. Baltimore, Md.: Johns Hopkins University Press for Resources for the Future, 1983.

Tuchmann, E. Thomas. "Statement of E. Thomas Tuchmann, Director, Resource Policy, Society of American Foresters, before the Subcommittee on Forests, Family Farms, and Energy, Committee on Agriculture, U.S. House of Representatives, on the Winding Stair National Recreation and Wilderness Area Bill." Bethesda, Md.: Society of American Foresters, 1988.

Twight, Ben W. *Organizational Values and Political Power: The Forest Service versus the Olympic National Park.* University Park: Pennsylvania State University Press, 1983.

Umbeck, John. "The California Gold Rush: A Study of Emerging Property Rights." *Explorations in Economic History* 14 (1977): 197–226.

———. "Might Makes Rights: A Theory of the Formation and Initial Distribution of Property Rights." *Economic Inquiry* 19 (1981): 38–59.

———. "A Theory of Contract Choice and the California Gold Rush." *Journal of Law and Economics* 20 (1977): 421–37.

———. *A Theory of Property Rights, with Application to the California Gold Rush.* Ames: Iowa State University Press, 1981.

"U.S. Reports Discovery of Minerals in Alaska." *New York Times,* September 3, 1975, p. 40.

U.S. Congress. House. "Report No. 968: Forest Reserves Administration, Minority Report." 57th Congress, 1st sess., 1901.

———. "Report No. 2521: Providing for the Preservation of Wilderness Areas." 87th Congress, 2d sess., 1962.

U.S. Congress. House. Committee on Interior and Insular Affairs. *Report to Accompany National Land Policy, Planning and Management Act of 1972.* 92d Congress, 2d sess., 1972.

U.S. Congress. House. Subcommittee on Mines and Mining of the Committee on Interior and Insular Affairs. *Hearings on National Minerals Security Act.* 97th Congress, 1st sess., 1981.

———. *Oversight Hearings on Nonfuel Minerals Policy Review, Parts I and II.* 96th Congress, 1st sess., 1979.

———. *Report on U.S. Minerals Vulnerability: National Policy Implications.* 96th Congress, 2d sess., 1980.

U.S. Congress. House. Subcommittee on Public Lands of the Committee on Interior and Insular Affairs. *Hearings on Public Land Policy and Management Act of 1975.* 94th Congress, 1st sess., 1975.

——. *Hearings on a Wilderness Preservation System.* 87th Congress, 1st sess., 1961.

——. *Hearings on a Wilderness Preservation System.* 87th Congress, 2d sess., 1962.

——. *Hearings on a Wilderness Preservation System.* 88th Congress, 2d sess., 1964a.

——. *Hearings on a Wilderness Preservation System (H.R. 9070, H.R. 9162, and S. 4).* 88th Congress, 2d sess., 1964b.

U.S. Congress. House. Subcommittee on Public Lands and National Parks of the Committee on Interior and Insular Affairs. *Oversight Hearings on Administration's Asset Management Program.* 98th Congress, 1st sess., 1983.

——. *Oversight Hearings on Public Land Management Policy.* 98th Congress, 2d sess., 1984.

——. *Oversight Hearings on Public Land Sales and Transfers.* 97th Congress, 2d sess., 1982.

U.S. Congress. House. Subcommittee on Science, Research, and Technology of the Committee on Science and Technology. *Hearings on a National Policy for Materials.* 95th Congress, 1st and 2d sess., 1978.

——. *Hearings on a National Policy for Materials.* 95th Congress, 2d sess., 1979.

U.S. Congress. House. Subcommittee on the Environment of the Committee on Interior and Insular Affairs. *Hearings on the Public Land Policy Act of 1971.* 92d Congress, 1st sess., 1971.

U.S. Congress. House. Subcommittee on Transportation, Aviation, and Materials of the Committee on Science and Technology. *Hearings on the National Critical Materials Act of 1984.* 99th Congress, 1st sess., 1985.

U.S. Congress. House. Subcommittees on Transportation, Aviation, and Materials and Science, Research, and Technology of the Committee on Science and Technology. *Oversight Hearings on PL 96-479 and Consideration of HR 4281, Critical Materials Act of 1981.* 97th Congress, 2d sess., 1982.

U.S. Congress. Senate. "Report No. 635: Establishing a National Wilderness Preservation System." 87th Congress, 1st sess., 1961.

U.S. Congress. Senate. Committee on Energy and Natural Resources. *Hearings on Federal Property Management and Disposal.* 97th Congress, 2d sess., 1982.

U.S. Congress. Senate. Committee on Interior and Insular Affairs. *Hearings on Legislation to Revise the Public Land Laws.* 92d Congress, 1st sess., 1971.

——. *Hearings on a Wilderness Preservation System.* 85th Congress, 1st sess., 1957.

——. *Hearings on a Wilderness Preservation System (S. 4028).* 85th Congress, 2d sess., 1958.

——. *Hearings on a Wilderness Preservation System (S. 1123).* 86th Congress, 1st sess., 1959.

——. *Hearings on a Wilderness Preservation System (S. 174).* 87th Congress, 1st sess., 1961.

——. *Hearings on a Wilderness Preservation System (S. 4).* 88th Congress, 1st sess., 1963.

——. "Public Land Policy: Activities in the 92nd Congress." 92d Congress, 2d sess., 1972.

U.S. Congress. Senate. Subcommittee on Energy and Mineral Resources of the

Committee on Energy and Natural Resources. *Hearing on the President's National Materials and Minerals Program and Report to Congress.* 97th Congress, 2d sess., 1982.

———. *Hearings on Strategic Minerals and Materials Policy.* 97th Congress, 1st sess., 1981.

U.S. Congress. Senate. Subcommittee on Energy Resources and Materials Production of the Committee on Energy and Natural Resources. *Hearings on Materials Policy, Research, and Development Act.* 96th Congress, 2d sess., 1980.

U.S. Congress. Senate. Subcommittee on Science, Technology, and Space of the Committee on Commerce, Science, and Transportation. *Hearings on Material Policy.* 95th Congress, 1st sess., 1977.

———. *Hearings on a National Materials Policy.* 96th Congress, 2d sess., 1980.

U.S. Council on Environmental Quality. *Environmental Quality 1984.* Washington, D.C.: Government Printing Office, 1985.

U.S. Department of Agriculture. *Annual Reports of the Department of Agriculture: 1900.* Washington, D.C.: Government Printing Office, 1900.

———. *Annual Reports of the Department of Agriculture: 1902.* Washington, D.C.: Government Printing Office, 1902.

———. *Annual Reports of the Department of Agriculture: 1904.* Washington, D.C.: Government Printing Office, 1904.

———. *Annual Reports of the Department of Agriculture: 1906.* Washington, D.C.: Government Printing Office, 1907.

———. *Annual Reports of the Department of Agriculture: 1907.* Washington, D.C.: Government Printing Office, 1908.

———. *Annual Reports of the Department of Agriculture: 1908.* Washington, D.C.: Government Printing Office, 1909.

U.S. Department of the Interior. Bureau of Land Management. *Public Land Statistics 1978.* Washington, D.C.: Government Printing Office, 1980.

———. *Public Land Statistics 1984.* Washington, D.C.: Government Printing Office, 1985.

———. *Public Land Statistics 1987.* Washington, D.C.: Government Printing Office, 1988.

———. *Public Land Statistics 1991.* Washington, D.C.: Government Printing Office, 1992.

U.S. Executive Office of the President. National Critical Materials Council. "A Critical Materials Report: The Continuation of a Presidential Commitment." Washington, D.C.: Government Printing Office, 1989.

U.S. General Accounting Office. *Numerous Issues Involved in Large-Scale Disposals and Sales of Federal Real Property.* Washington, D.C.: Government Printing Office, 1981.

U.S. Office of Management and Budget. *Budget of the United States, Fiscal Year 1988 (and Appendix).* Washington, D.C.: Government Printing Office, 1987.

U.S. Office of Technology Assessment. *Management of Fuel and Nonfuel Minerals in Federal Land: Current Status and Issues.* Washington, D.C.: Government Printing Office, 1979.

———. *Strategic Materials: Technologies to Reduce U.S. Import Vulnerability.* Washington, D.C.: Government Printing Office, 1985.

U.S. Secretary of the Interior and Secretary of Agriculture. *Study of Fees for Grazing Livestock on Federal Lands.* Washington, D.C.: Government Printing Office, 1977.

Van Hise, Charles R. *The Conservation of Natural Resources.* New York: Macmillan, 1910.

Voigt, William. *Public Grazing Lands.* New Brunswick, N.J.: Rutgers University Press, 1976.

Wade, Nicholas. "Raw Materials: U.S. Grows More Vulnerable to Third World Cartels." *Science* 183 (January 18, 1974): 185–86.

Watkins, T. H. "The Perils of Option 9." *Wilderness* (Winter 1993): 6–9.

"Watt Vows Shift on Key Minerals." *New York Times*, March 3, 1981, p. B10.

Weir, Margaret. *Politics and Jobs: The Boundaries of Employment Policy in the United States.* Princeton, N.J.: Princeton University Press, 1992.

Wiebe, Robert H. *The Search for Order, 1877–1920.* New York: Hill and Wang, 1967.

"Wilderness Watch: A 'Subsidy of Destruction' Brings Forth a Suit." *Wilderness* (Summer 1986): 2.

"Wilderness Watch: Grazing Subsidy Survives." *Wilderness* (Spring 1986): 4.

Wilkinson, Charles F. *Crossing the Next Meridian: Land, Water, and the Future of the West.* Washington, D.C.: Island Press, 1992.

Williamson, Lonnie. "A Range Is a Terrible Thing to Waste." *Outdoor Life* (October 1985): 30–33.

———. "Where the Grass Is Greenest." *Outdoor Life* (February 1985): 30–31.

Wilson, James Q. *Bureaucracy: What American Government Agencies Do and Why They Do It.* New York: Basic Books.

Winter, John R., and James K. Whittaker. "The Relationship between Private Ranchland Prices and Public-Land Grazing Permits." *Land Economics* 57 (1981): 414–21.

Wood, Gordon S. "Rhetoric and Reality in the American Revolution." *William and Mary Quarterly* 23 (1966): 3–32.

Yaffe, Steven L. *The Wisdom of the Spotted Owl: Policy Lessons for a New Century.* Washington, D.C.: Island Press, 1994.

Yard, Robert S. *Our Federal Lands.* New York: Scribner's, 1928.

Zaleha, D. Bernard. "The Rise and Fall of BLM's 'Cooperative Management Agreements': A Livestock Management Tool Succumbs to Judicial Scrutiny." *Environmental Law* 17 (1986): 125–52.

INDEX

National Park Service (NPS), 5, 22, 77

National recreation areas (NRAS), 55

National security. *See* Mining policy: and national security

National Strategic Materials and Minerals Program Advisory Committee, 61

National Wilderness Preservation Council, 81, 89

National Wilderness Preservation System, 38

National Wildlife Federation, 60, 134

National Wool Grower's Association, 131–32

Natural Resources Defense Council (NRDC), 60, 134

Nature: preservationism and value of, 20–25. *See also* Preservationism

New Mexico Cattle Growers' Association, 3

New Resource Economics (NRE), 95–96

Nicholson Report, 127

Nixon, Richard, 118–19

Northern Rockies Ecosystem Protection Act (NREPA), 152. *See also* Montana wilderness

Northwest Mining Association, 41

Office of Management and Budget (OMB), 96, 97, 134, 138

Office of Technology Assessment (OTA), 49

Oil and gas development, 57–58, 62, 63, 167 (n. 23)

Oil embargo. *See* Mineral policy: minerals crisis and shortages

One Third of the Nation's Land. See Public Land Law Review Commission

Oregon Endangered Species Task Force. *See* Spotted owl dispute

Organic act, Bureau of Land Management (BLM). *See* Federal Land Policy and Management Act

Overton, J. Allen, 50, 52

Pinchot, Gifford, 15–17, 71, 73, 75. *See also* Forest Service

Pluralist gridlock, 144, 159

Policy patterns, 5–9, 141–44, 161 (nn. 10, 11); captured (grazing), 109, 115, 126, 140, 144–45, 157; privatized (mining), 27, 35–36, 64–66, 143–44; professional (forestry), 63, 99, 107, 143. *See also* Forestry profession; Forest Service; Grazing policy; Mining policy

Policy regimes. *See* Forestry profession; Forest Service; Grazing policy; Mining policy

Polk, James, 30

Preservationism, 20–26, 175 (n. 21); and radical groups, 151–53. *See also* Environmentalists; Preservationists

Preservationists: and forestry policy, 37–38, 76, 142; and grazing policy, 120–21, 130, 133, 139–40, 142; and mining policy, 27–28, 31, 37–38, 47, 65, 142; and wilderness regulation, 23, 51, 78–81, 142, 151. *See also* Environmentalists; Preservationism; Wilderness Act

Private ownership, 12–13. *See also* Privatization movement

Privatization movement, 93–107, 176 (n. 55), 177 (n. 65). *See also* Economic liberalism

Property Review Board, 97–98, 106

Property rights, 114, 130, 167 (n. 22); extralegal, 29, 32, 35–36, 39, 64, 93, 110, 140, 145. *See also* Grazing policy; Mining policy

Public domain lands. *See* Public lands

Public Land Law Review Commission (PLLRC), 13, 179 (n. 20); *One Third of the Nation's Land*, 116–17

Public lands: management reform of, 1–5, 116–26; privatization of, 13, 34, 69, 111, 143; transfer to states, 12, 14, 94, 111, 176 (n. 53). *See also* Sagebrush Rebellion

ownership, ideas about; Sustainable development
Technocrats. *See* Technocratic utilitarianism
Thomas, Jack Ward, 149–50
Thoreau, Henry David, 20
Timber industry and interests, 13, 79, 84, 87, 90–91, 155; clear cutting, 92, 156–57. *See also* Spotted owl dispute
Transcendentalism, 20, 24

Udall, Morris, 118, 136
U-Regulations. *See* Forest Service: and administrative wilderness system
U.S. army/military. *See* Mining policy: U.S. army/military
U.S. Forest Service. *See* Forest Service
U.S. Geological Survey (USGS), 36–37, 46, 56

Wallop, Malcolm, 2, 137
Watt, James, 54, 57, 59–61, 94, 103
Western Livestock Grazing Survey, 129
Western States Governors Conference, 13
Wild and Scenic Rivers Act of 1968, 23

Wilderness Act, 23, 26; designation and expansion of, 48, 87, 89, 92, 124; fires, insects, and disease control, 88, 91; general debate over legislation, 76–93; and Indian lands, 80–81, 89, 91; and mining exemption, 37–47, 48, 50, 51, 53, 55–62, 64, 82, 84–85, 91, 92, 168 (n. 40), 169 (nn. 45, 48), 171 (n. 31); and motorized recreation, 78, 80, 84, 91, 147; and water development, 82, 84, 86, 88, 91, 92. *See also* Environmentalists; Federal Land Policy and Management Act: and wilderness; Forest Service: and administrative wilderness system; Preservationists
Wilderness Protection Act, 57–59
Wilderness Society, 22, 24–26, 60, 78, 120–21. *See also* Alaska Coalition
Wildlife management, 23–24, 80–82, 88, 155. *See also* Spotted owl dispute
Wise use movement, 15, 147–48. *See also* Interest-group liberalism

Yellowstone National Park, 21, 77

Zahniser, Howard, 23, 38, 45, 79–80